Loren Eiseley

Commentary, Biography, and Remembrance

Edited by Hilda Raz
Introduction by Scott Slovic

UNIVERSITY OF NEBRASKA PRESS
LINCOLN AND LONDON

∞

The essays in this volume originally appeared
in *Prairie Schooner* 61, no. 3 (Fall 1987). © 1987
by the University of Nebraska Press.

"Loren Eiseley in Lincoln: Two Poems and a
Remembrance" reprinted by permission of
Gale E. Christianson.

Library of Congress Cataloging-in-
Publication Data
Loren Eiseley: commentary, biography,
and remembrance / edited by Hilda Raz;
introduction by Scott Slovic.
    p. cm.
"The essays in this volume originally
appeared in Prairie Schooner 61, no. 3."
Includes bibliographical references.
ISBN 978-0-8032-1906-9 (pbk.: alk. paper)
1. Eiseley, Loren C., 1907–1977. 2. Eiseley,
Loren C., 1907–1977—Criticism and
interpretation. 3. Authors, American—20th
century—Biography. 4. Anthropologists—
United States—Biography. 5. Naturalists—
United States—Biography. I. Raz, Hilda.
II. Prairie schooner.
PS3555.I78Z695   2008
818.5′409—dc22
2008029071

The essays in this collection follow their
original format by beginning on page 5. The
text remains unaltered.

CONTENTS

Introduction

The eerie fluorescent brightness of the airport lounge was accented by the *Spirit of St. Louis*, hanging pilotless from the rafters. I was twenty-three years old and had just embarked on a week-long tour of graduate programs, hoping to learn enough during brief visits to half a dozen English departments to make a decision about where to spend the next five years of my life. I arrived at the airport in St. Louis, Missouri, close to midnight on the last flight of the day. Having made no plans for the night, I spent six hours dozing and hallucinating in the empty waiting area near the ghostly replica of the plane Charles Lindbergh had flown alone for thirty-three hours across the Atlantic in 1927.

At dawn, sleepless and red-eyed, I caught a cab over to the campus of Washington University, figuring I could find a bench to sprawl out on until the department office opened. My memory is dim now, decades later, but I believe I simply walked past the English department, not pausing to knock, when a door opened. It was 6:00 a.m., an hour when few professors are in their offices. An elderly man with a crew cut peered out and asked, "Are you Scott Slovic? We've been expecting you." "I'm Howard Nemerov," the man nervously introduced himself. He escorted me to a room where I could wait for the department secretary and then disappeared. Our encounter lasted all of two minutes, leaving me with the memory of his quavering, anxious-sounding voice more than anything else. I never saw Professor Nemerov again, but I've often wondered in the years since whether his early arrival at the English department was a daily occurrence, whether he frequently anticipated prospective graduate students in such a way, or whether he may have taken special interest in my application to the department because I had recently completed an undergraduate thesis devoted, in part, to the work of Loren Eiseley.

Readers of this volume (a reprinting of part of the special fall 1987 *Prairie Schooner*) will note that Howard Nemerov wrote the opening essay, "Loren Eiseley 1907–1977." Nemerov, the great poet, was an Eiseley fan. He deferentially approached Eiseley in a June 1971 letter, thanking him for "the solitude" in his books, which Nemerov found "strong and oddly consoling." Eiseley responded two weeks later with a letter inviting Nemerov to meet him for lunch if he ever came to Philadelphia, so they could tell each other "sad stories." Only a few letters between the two are now housed in the Gale E. Christianson Collection of Eiseley Research Materials at Bennett Martin Public Library in Lincoln, Nebraska. But it's clear the two established a quick rapport, even exchanging humorous anecdotes (initiated by Eiseley) about being pulled out of line various times by airport security in 1973, a time of post-Munich heightened sensitivity to terrorism oddly similar to our own. When Nemerov introduced the special issue of *Prairie Schooner* in 1987, ten years after his friend's death, he offered a poignant assessment of Eiseley's importance to literary readers such as himself:

> I suppose that what we poor literary folk want in the first place is not to be talked to *de haut en bas* by Lord Snow, and certainly not in the second place to be defended by Dr. Leavis, but to be written to on scientific subjects with eloquence, good sense, patience, loneliness rather like our own, and a strong, continuous awareness of the presence of the knower in the known; which is what I have found in a small number of physicists and biologists, and what I derive from Loren, paleontologist, anthropologist, "bone-hunting man" as he described himself, saying "I am a man who has spent a great deal of his life on his knees, though not in prayer," and, finally, poet. (7)

What Nemerov, like many others, loved about Eiseley's work was the sense of passionate investment in literary style and the moments of authorial connection with other beings. He loved the way even Eiseley's ordinary experiences carried high meaning and were attached to correspondingly elevated language, the way the author expressed his subtle (and almost embarrassing) sense of kinship with other solitary beings (birds, bones, and fellow wanderers), and the secular prayerfulness of Eiseley's sensibility (that of a man who enacted devotion to the world through hands-on engagement with nature rather than through indoor displays of piety). I especially resonate to Nemerov's lovely and gently paradoxical line "loneliness rather like our own," for it hints at the special bond of friendship he shared with Eiseley, as well as recognizes the cosmic solitude that is so effectively (and sympathetically) described in Eiseley's work. The memory of my own brief encounter with Nemerov stays with me through the years. His early morning loneliness heartens me, a reminder of the loneliness we all share.

In addition to the opening essay by Nemerov, the fall 1987 issue of *Prairie Schooner* contains several compelling gems. Gale E. Christianson, the author of *Fox at the Wood's Edge: A Biography of Loren Eiseley* (1990), condensed snippets from his biographical study into "Loren Eiseley in Lincoln: Two Poems and a Remembrance," providing rare glimpses into young Eiseley's social life and early literary influences. Although some of this material also appears in the biographical tome, this smaller essay is impressive in its description of Eiseley's physical resilience, his attractiveness to women, and his early engagement with the ideas and expressive styles of evolutionist Thomas H. Huxley and poet Robinson Jeffers. Erleen J. Christensen explores Eiseley's intertwined fascinations with time and mortality in her examination of his poetry and prose "Loren Eiseley, Student of Time." The author of *Loren Eiseley: The Development of a Writer* (1983), E. Fred Carlisle, provides a succinct synopsis of Eiseley's scientific work and literary projects, emphasizing the importance of the author's "dual identity" as scientist and writer in his distinctive artistic achievement. Peter Heidtmann's *Loren Eiseley: A Modern Ishmael* appeared in 1991, exploring the writer's psychological, philosophical, and autobiographical tendencies; his contribution to the special issue of *Prairie Schooner*, "An Artist of Autumn," argues that Eiseley crafted a persona imbued with morose "autumnal consciousness," one that resonates powerfully with readers. In "Loren Eiseley and the State of Grace," Ben Howard highlights a very different aspect of Eiseley: his unique "gift for evoking, in one or two pages, the spiritual condition of man in a state of grace." Eiseley's personal assistant, Caroline E. Werkley, herself a fey believer in magic and spirits, argues in "Eiseley and Enchantment" that if Eiseley had lived in primitive times, he himself would have been a shaman, priest, or magician, a "provider of medical, magical and religious guidance." Social worker Naomi Brill, in "Loren Eiseley and the Human Condition," discerns a complex and suffering humanity in

Eiseley-the-man and Eiseley-the-artist, suggesting that his artistic achievements are inextricable from the "burden of terror, sadness, anger and guilt [he carried] throughout his life." The contributors to this special volume, from Nemerov to Brill, represent a who's who of Eiseley scholars, and it is a great pleasure to make their work available to a new generation of readers.

Some three decades following Loren Eiseley's death—and a century after his birth—Eiseley remains a prominent figure in the tradition of American literature, particularly within the genres of literary nonfiction, autobiography, and environmental writing. My own initial fascination with his work, back when I was a student of autobiography at Reed College and Stanford University, developed as a result of my chance discovery of a copy of *All the Strange Hours: The Excavation of a Life* (1975) at a used bookstore in Madison, Wisconsin, during a cross-country bus trip from California to Pennsylvania, where I stayed for several months with my uncle in West Philly, roaming the neighborhoods where Eiseley spent much of his literary career. Widely reviewed and praised, *All the Strange Hours* has come to be regarded as a major contribution to American autobiographical writing, but as I read more and more of Eiseley's literary work—especially his many collections of personal essays, which he himself referred to as "concealed essays" (*Strange Hours* 177)—my own feeling was that *all* of Eiseley's writing, in one way or another, was part of a subtle (perhaps even unconscious) effort of self-discovery. As Nemerov obliquely commented, Eiseley wrote on "scientific subjects" with "continuous awareness of the presence of the knower in the known," a statement I take as a reference to the fact that Eiseley saw *himself* in the far-flung paleontological, evolutionary, and history-of-science subjects he explored in his work.

For various reasons, scholars in the field of ecocriticism (ecological literary criticism) regard Eiseley as a pivotal figure in mid-twentieth-century American environmental literature. Don Scheese, in *Nature Writing: The Pastoral Impulse in America* (1995), notes that for Eiseley the natural world was not relegated to the distant wilderness but existed in close relation to human experience everywhere humans happened to be, "in locales as multiform as the Badlands of the Dakotas and New York City." According to Scheese, Eiseley "helped popularize appreciation of nature by recognizing it as a cultural universal," perhaps because of the author's training in anthropology and his understanding of cultures beyond the narrow tradition of Anglo-American nature writing. Although he doesn't refer explicitly to Eiseley's incessant fascination—indeed, obsession—with the *experience* of evolutionary process, Scheese hints at the importance of this aspect of Eiseley's work in commenting that the author "reminds readers . . . of the connections among all living things" (31). One of the central dilemmas of environmental literature, as contemporary writer Scott Russell Sanders has stated eloquently, is "how, despite the perfection of our technological boxes, to make us feel the ache and tug of that organic web passing through us, how to *situate* the lives of characters—and therefore of readers—in nature" (226). Situating our lives in nature and nudging readers to feel viscerally the tug of this "organic web" as it passes through their very bodies is precisely the object of Eiseley's work, I believe. When Eiseley himself looked from his office window at the University of Pennsylvania and sensed the imminent emergence of new life in a rooftop puddle, he was extending himself beyond the "technological box" of his building, participating consciously and emotionally in the planet's ecology.

Barry Lopez and others have explained that so-called nature writing is sometimes better described as "the literature of place." Lopez wrote in a 1998 essay, "A specific and particular setting for human experience and endeavor is, indeed, central to the work of many nature writers. I would say, further, that it is also crucial to the development of a sense of morality and human identity" (1). It is certainly true that Eiseley evoked various places in rich and compelling prose—one thinks of numerous renderings of his native Great Plains and his adopted eastern cityscapes and also of such locales as coastal Florida, memorably presented in "The Star Thrower." But, as ecocritic Fred Taylor has explained, Eiseley's "descriptions of landscape . . . are pervaded by a sense of mystery. . . . [W]ith Eiseley, we often do not even know where we are." Instead, Taylor continues, "in directing our attention away from the specificity of a given setting, he seems more interested in engaging us in the feel of a particular moment of awareness or encounter. Such epiphanies are frequent in Eiseley's writing, often coming at the climax of a long search that prepares him for a revelation" (274).

It is certainly reasonable to characterize Eiseley's work as a contribution to "the literature of place," but one must hasten to observe that for Eiseley the specificities of place were significant mostly to the extent that they play a role in shaping the human subject's state of mind. Place matters but chiefly as a locus of experience—of revelation. One vivid example of this comes from the splendid essay "The Bird and the Machine," collected in *The Immense Journey* (1957), which remains the best-known (and most frequently taught) volume of Eiseley's essays. Early in the essay, the author establishes the setting as some remote and nameless high plains landscape, referring to "the stones and winds of those high glacial pastures" and "the mountain snows" (178). He hints at his solitary condition and his inadvertent yearning for companionship of any kind when he recalls sitting "on a high ridge that fell away before [him] into a waste of sand dunes," next to a large, coiled rattlesnake: "We were both locked in the sleep-walking tempo of the earlier world, baking in the same high air and sunshine" (184). The scene has been set. This is a remote, lonely place, a place so lonely that a man—such as Eiseley—would appreciate the company of a sluggish rattler, both retreating mysteriously to the evolutionary status of "an earlier world," a time when their kinship as living beings would have been obvious. It is here, in such a place and such a state of mind, that Eiseley becomes able to achieve what Lopez calls "a sense of morality and human identity." Eiseley's identity stretches beyond the merely human. At the narrative's climax, he understands that a captured sparrow hawk has a mate who has been circling overhead throughout the night, hoping to be reunited. Without quite letting the meaning of his actions "come up into consciousness," Eiseley lifts the male hawk from the box, "feel[ing] his heart pound under the feathers," sensing his essential life-energy, his similarity to the living, feeling author (191). At this moment, the revelation occurs:

> I was young then and had seen little of the world, but when I heard that cry my heart turned over. It was not the cry of the hawk I had captured; for, by shifting my position against the sun, I was now seeing further up. Straight out of the sun's eye, where she must have been soaring restlessly above us for untold hours, hurtled his mate. And from far up, ringing from peak to peak of the summits over us, came a cry of such unutterable and ecstatic joy that it sounds down across the years and tingles among the cups on my quiet breakfast table. (191–92)

The moral lesson learned through this encounter is that all living beings bear something in common, something that makes them different from machines. The essay, published in the

1950s, appeared at the dawn of the computer era, and the stated occasion for this memory of the release of the bird many years before is the author's scanning of the newspaper over breakfast and recoiling at the news of scientists' efforts to create "smart," lifelike technologies. For Eiseley, there was something unique and magical about life itself, something essentially beyond what humans can manufacture. His writing is a way of tapping into that magic, reveling in the feeling of shared vitality and moral actions in support of life, ranging from the release of a captured bird to the flinging of stranded starfish back into the surf.

Eiseley's literary nature writing and his scholarly forays into the history of science overlap in the author's abiding devotion to the idea and the experience of evolution. His many studies of the history of evolutionary thought, such as his early book *Darwin's Century: Evolution and the Men Who Discovered It* (1958), helped to clarify the elegant rationality of evolution, its power as a synthetic explanation of human nature and the nature of life in general. For Eiseley, to study evolution was a means of placing humankind in *time*, of ferreting out the clues to human history and anticipating the future of human development; it was also a way of living intensely in the present, celebrating the constant flux of being. Again and again in Eiseley's work, we find ourselves transfixed by stories plucked from quotidian oblivion by the writer's imagination and illuminated as opportunities to appreciate the mystery of life and the possibility of achieving community, of overcoming cosmic loneliness. Eiseley's essay "The Flow of the River," also from *The Immense Journey*, offers exquisite examples of this emphasis on *mystery* and *community*. The essay's opening passage, for instance, suggests famously, "If there is magic on this planet, it is contained in water." Even in an ordinary puddle on a Philadelphia rooftop following a rainstorm, "a wind ripple may be translating itself into life" (15). The essay concludes in a vein that stops just short of being explicitly religious: "There is no logical reason for the existence of a snowflake any more than there is for evolution. It is an apparition from that mysterious shadow world beyond nature" (27). Though Eiseley does not use the word "God," it seems clear that he was open to the possibility of nature's numinous qualities, to the ineffabilities that exceed human ratiocination.

I have already suggested above that the solitary Eiseley repeatedly sought moments of connection with fellow solitaries, ranging from authors such as Howard Nemerov to snakes, birds, ancient skulls newly extracted from the earth, and life-giving water itself. In "The Flow of the River" he writes of one occasion when he allowed himself to overcome his childhood fear of drowning in order to float downstream in the Platte River, experiencing a great joining with the universe:

> I lay back in the floating position that left my face to the sky, and shoved off. The sky wheeled over me. For an instant, as I bobbed into the main channel, I had the sensation of sliding down the vast tilted face of the continent. It was then that I felt the cold needles of the alpine springs at my fingertips, and the warmth of the Gulf pulling me southward. . . . I was streaming over ancient sea beds thrust aloft where giant reptiles had once sported; I was wearing down the face of time and trundling cloud-wreathed ranges into oblivion. I touched my margins with the delicacy of a crayfish's antennae, and felt great fishes glide about their work. (18–19)

When Eiseley writes that he "touched [his] margins," he suggests that he felt the essential shape, the precise borders, of his being, understanding his connectedness to the earth and all the lives it supports, past, present, and future. This visionary sense of "community" complements the many episodes in Eiseley's work in which he describes his eccentric communitarian impulses. Prompted by the phrase "eccentric community," many readers will immediately think of the famous episode in "The Judgment of the Birds," also from *The Immense Journey*, in which the insomniac Eiseley, awake and lonely in a New York City hotel room in the middle of the night, feels himself momentarily to be a bird and nearly leaps from his hotel window: "It needed only a little courage, only a little shove from the window ledge to enter that city of light. The muscles of my hands were already making little premonitory lunges" (166). Other readers will remember Eiseley's character Albert Dreyer from the short story "The Dance of the Frogs" (in *The Star Thrower* [1978]), who is so stirred by the springtime rhythm of frogs hopping from hibernation back toward the newly thawed river that he begins to hop along with them, one of his hands actually reverting atavistically to a froglike web. "Part of me struggled to stop," he states in Eiseley's story, "and part of me hurtled on" (114). One senses this very ambivalence in Eiseley's own identity, his sense of fundamental inbetweenness—between science and literature, between the rural plains and the eastern city, between human and nonhuman, between past and future. And from this essential state of alienation and loneliness comes the extravagant yearning to be connected.

Perhaps the most unique and disturbingly memorable aspect of Eiseley's evolutionary narratives is the representation of evolution as an ongoing, immediate phenomenon, something any one of us could witness in our daily lives if we peer hard enough, if we are willing to risk genuine perception of the fluctuating, insubstantial condition of life. One of the mantras of *All the Strange Hours* is the cryptic phrase "behind nothing / before nothing / worship it the zero," which Eiseley attributes vaguely to "someone" (6). "The zero" implies the self and the present moment. And indeed, as Eiseley relates his own experiences in the autobiography *All the Strange Hours,* we encounter with him such uncanny physical specimens as Tim Riley, his traveling companion on a train, whose fingernails "were raised and thickened like the claws of an animal. Furthermore, they ended as sharply as claws. Tiger claws, I think with a shudder, except that they were not retractile" (118). Eiseley explains this observation as an example of the unique perception of a "physical anthropologist," who, "perhaps alone, is the most conscious of human differences, strange mutations in the normal run of things, inexplicable emergencies, atavisms, all that difficult entangled thread that produces successive generations" (120). I would argue, though, that Eiseley's repeated encounters with bodies-in-flux, beings symbolizing evolutionary inbetweenness, represent the author's fundamental belief that evolution is observable in the present moment, the *now.*

Assessing Eiseley's importance in the tradition of American literary nonfiction (and, more specifically, the tradition of nature writing), Annie Dillard wrote in 1988, "In this century, it was Loren Eiseley—a scientist—who restored the essay's place in imaginative literature and who extended its symbolic capacity. . . . Eiseley lays in symbols with a trowel, splashing mortar right and left—but the symbols hold" (xv). Herself a playful and flamboyant inventor of symbols, Dillard insightfully captures one of the essential features of Eiseley's essayistic mode: his constant, simultaneous representation of experience in both vividly realistic and mind-blowingly symbolic terms. His detailed narratives, rich with dialogue and introspection, compel us to believe he is telling us "the truth." And his unlikely characters and profound interpretations of experience make the symbolic meaning of his reports unmistakable. Perhaps one of the litmus tests for readers of

Eiseley's works, a separation point for enthusiasts and detractors, is the issue of how explicit and explanatory we expect an author to be. If we like our literature murky and oblique, Eiseley might not be our cup of tea. If we like earnest clarity, an almost desperate yearning to communicate, Eiseley will hit the spot.

As I mentioned above, Eiseley's single best-known book is certainly *The Immense Journey*. Many of its essays have been collected in anthologies, and generations of American high school and university students have learned to write by emulating the essay style exhibited in its pages: illustrative anecdotes interspersed with philosophical commentary. However, for scholars interested in the ongoing development of evolutionary theory, Eiseley's take on the subject in this 1957 collection is likely to seem unsophisticated and dated, less compelling than the more recent work of Stephen Jay Gould, Richard Dawkins, and other contemporary expositors.

Readers of Eiseley's work may find his assertions of political indifference (or outright political conservatism) to be uninspiring. In *All the Strange Hours* he describes the irony of the fact that he became a hot commodity as a public speaker during the era of the Vietnam War—that he became a public figure at *all*. He notes, "I had emerged as a rather shy, introverted lad, to exhort others from a platform. . . . For me it has been a lifelong battle with anxiety" (132). He goes on in this passage to shake his head at the apparent enthusiasm for his apolitical topics during the superheated political atmosphere of the 1960s:

> I am not a political speaker but for some obscure reason I was occasionally to find myself on violent campuses delivering lectures on such subjects as "Ice, Time, and Human Destiny" just at a period when some new bewhiskered idol was delivering a competitive address on smut, or extolling the virtues of fire-bombing the whole place into oblivion. I have never quite understood why my recondite subject matter drew the audiences it did, unless the campuses were so hot that, as in the case of the title just quoted, even students wanted a breath of cool air off the ice sheet. (132)

Never a "bewhiskered idol" himself, Eiseley resisted the general political tide on college campuses during the Vietnam era. When I visited the Eiseley Archive at the Bennett Martin

Public Library in 2007, I chanced upon a brief correspondence between anthropologist Margaret Mead, writing on behalf of the American Association for the Advancement of Science (AAAS), and Eiseley, then provost at the University of Pennsylvania, on the subject of national security versus freedom of scientific information. Mead had written to Eiseley to feel him out on this issue during the Vietnam War, presumably expecting Eiseley to side with his scientific colleagues on the importance of combating secret science, but Eiseley instead expressed sympathy for the government's need to classify certain scientific discoveries in the interest of wartime national security.

Despite his conservative stances on social and military issues, Eiseley did express his progressive environmental concerns quite eloquently in his later publications, including *The Invisible Pyramid* (1970) and *All the Strange Hours*. Just a few pages after disavowing politics in the autobiography, for instance, he writes, "Like the ice, [mankind has] been cruel to the face of the planet and the life upon it. A chill wind lingers about us. With a few slight exceptions we are merciless. We have invented giant, earth-scavenging machinery to do what the ice once did" (155). This passage, couched in the terms of Eiseleyan Ice Age paleontology, clearly echoes Aldo Leopold's classic lines from *A Sand County Almanac* (1949): "We are remodeling the Alhambra with a steam-shovel, and are proud of our yardage. We shall hardly relinquish the shovel, which after all has many good points, but we are in need of gentler and more objective criteria for its successful use" (225–26).

In *The Invisible Pyramid*, responding to the dawn of contemporary ecological consciousness in America, Eiseley writes bleakly—and with characteristically vivid metaphors—of the destructive role industrial civilization has played on the earth. "Perhaps man," he writes, "like the blight descending on a fruit, is by nature a parasite, a spore bearer, a world eater" (53). Echoing Paul Ehrlich's *The Population Bomb* (1968), Eiseley expresses alarm at human overpopulation: "The urban mass is reaching fantastic numbers and . . . the birthrate must be reduced" (108). He detours in a chapter called "The Time Effacers" from his environmentalist critique of industrial civilization to complain about what he perceives as the mindless, memoryless, and undisciplined contemporary youth (remember, the embedding of humanity in time is one of the central concerns of Eiseley's evolutionary work): "With the destruction of emotional continuity between the generations, time past is vilified or extinguished in favor of oncoming uncoordinated activist time. . . . The LSD trip has reached a level with the experience of the classroom. It is not a coincidence that the memory effacers have emerged in the swarming time of the spore cities" (109).

Here it sounds almost as if Eiseley is blaming the ills of urban sprawl ("spore cities") and the population explosion on drug-besotted, sex-crazed young people, not on real-estate developers, religious mandates to go forth and make babies, and the metasta-sizing corporate machine. Whatever the source of the dangerous fertility of mankind, the childless Eiseley writes with jeremiadic severity about the urban sprawl that "resembles a fungus upon a fruit": "Not long ago I chanced to fly over a forested section of country which, in my youth, was still an unfrequented wilder-ness. Across it now suburbia was spreading. . . . From my remote, abstract position in the clouds I could gaze upon all below and watch the incipient illness as it spread with all its slimy tendrils through the watershed" (161).

Of course, the sprawl Eiseley observed in the late 1960s has only continued during the past four decades, making his concerns both prescient and painful for today's readers and raising questions about the efficacy of activist writing. Perhaps, though, the power of such writing is not in its immediate effectiveness in producing social change but in its ability to inspire and sustain those who would seek to rage against the machine. The literature of social—and environmental—activism is not a blueprint for change but a beacon of hope.

The final point I would like to make about Eiseley's continuing significance for readers in the twenty-first century is that much of his writing—from the early evolutionary meditations to the later environmental and autobiographical works—is an eloquent expression of "hope in the dark," to use the phrase of contemporary essayist-activist Rebecca Solnit. *The Invisible Pyramid* was published during the space race, which Eiseley interpreted as an almost desperate response to news of the environmental crisis on this planet: "Was it fear of his own mounting numbers, the creeping of the fungus threads? But where, then, did these men intend to flee?" (153). Rather than thinking of alternatives to life on Earth, Eiseley viewed the 1968 photo taken from the Apollo 8 spacecraft of the earth rising above the lunar horizon, read a fearful magazine article predicting the end of human life on our home planet, and responded, "No, . . . this planet nourished man" (153).

For all of his brooding idiosyncrasies, Eiseley included among his own attributes (along with "Platonist," "mystic," and "concealed Christian") the phrase "midnight optimist" (*The Innocent Assassins* 11). Raised in a strained and unhappy household, victim of tuberculosis as a university student and deafness as a professor, witness to two world wars and other demoralizing examples of inhumanity (such as the Vietnam War), Eiseley nonetheless sought through his literary work to give voice to the world's magic, magic discerned in the most ordinary of circumstances. One might even argue that it was the difficulty of the author's life and the century during which he lived—its very *darkness*—that inspired him to turn from science toward the truths available to the literary imagination, toward the truths of experience. As he writes in "The Mind as Nature," "I, who endured the solitude of an ice age in my youth, remember now the yellow buttercups of the only picnic I was ever taken on in kindergarten. There were other truths than those contained in laboratory burners, on blackboards, or in test tubes. . . . I am trying to write honestly from my own experience" (*Night Country* 223).

Along with the honest reports of darkness come the honest reports of yellow buttercups. A more scientific analysis of the dominant trends of Eiseley's life and historical era might emphasize the bloody decline of civilization during seven decades of war, social strife, and environmental degradation. But the emotional truth of his life, accounted for in his literary work, includes many moments of humor and friendship and beauty. This supple combination of light and dark may be what continues to draw readers to Eiseley and to the writers—such as Rick Bass, David G. Campbell, Alison Hawthorne Deming, John Daniel, Pam Houston, Barbara Kingsolver, David Mas Masumoto, Gary Paul Nabhan, Richard K. Nelson, Robert Michael Pyle, Janisse Ray, Terry Tempest Williams, and many others—who have picked up where he left off.

One of the most memorable images Eiseley left us is that of a human being walking along a desolate shoreline—not just a literal beach but the "long resounding shores of endless disillusionment," the very symbol of our emotional condition in the modern world. This human walker, called "I" in the story/essay, encounters another character called simply "the star thrower," who is engaged in an activity at once futile and inspiring: tossing living creatures back from the beaches of death into their ocean home. Eiseley reminds us that even in times and places of desolation, there "emerges the awesome freedom to choose" either to dwell upon despair or to act hopefully. He writes, "In the sweet rain-swept morning, that great many-hued rainbow still lurked and wavered tentatively behind him. Silently I sought and picked up a still-living star, spinning it far out into the waves. I spoke once briefly. 'I understand,' I said. 'Call me another thrower.' Only then I allowed myself to think, He is not alone any longer. After us there will be others" (*Star Thrower* 88–89).

The world's beauties are "tentative," "lurking," potentially beyond our emotional reach. We are susceptible to despair. But despite the forceful momentum of our evolution and despite the genuine mess of a world we inhabit, we still possess the "freedom to choose" what perspective we'll adopt and what actions we'll take. Loren Eiseley, the scientific realist and mystical artist, opted in his most enduring work to be a "midnight optimist." As I survey the American literary landscape today, I think to myself, "He is not alone any longer." And I hope that after us "there will be others." Fellow travelers.

Works Cited

Carlisle, E. Fred. *Loren Eiseley: The Development of a Writer*. Urbana: U of Illinois P, 1983.

Christiansen, Gale E. *Fox at the Wood's Edge: A Biography of Loren Eiseley*. New York: Holt, 1990.

Dillard, Annie, ed. *The Best American Essays 1988*. New York: Ticknor & Fields, 1988.

Ehrlich, Paul. *The Population Bomb*. New York: Ballantine, 1968.

Eiseley, Loren. *All the Strange Hours: The Excavation of a Life*. New York: Scribner's, 1975.

———. *Darwin's Century: Evolution and the Men Who Discovered It*. Garden City NY: Doubleday, 1958.

———. *The Immense Journey*. New York: Random House, 1957.

———. *The Innocent Assassins*. New York: Scribner's, 1973.

———. *The Invisible Pyramid*. New York: Scribner's, 1970.

———. *The Night Country*. New York: Scribner's, 1971.

———. *The Star Thrower*. New York: Harcourt Brace Jovanovich, 1978.

Heidtmann, Peter. *Loren Eiseley: A Modern Ishmael*. Hamdon CT: Archon, 1991.

Leopold, Aldo. *A Sand County Almanac*. New York: Oxford UP, 1949.

Lopez, Barry. "We Are Shaped by the Sound of Wind, the Slant of Sunlight." *High Country News* 14 Sept. 1998: 1+.

Nemerov, Howard. Letter to Loren Eiseley. 5 June 1971. Gale E. Christianson Collection, Bennett Martin Public Library, Lincoln NE.

———. "Loren Eiseley 1907–1977." *Prairie Schooner* 61.3 (Fall 1987): 5–8.

Sanders, Scott Russell. *Secrets of the Universe: Scenes from the Journey Home.* 1987. Boston: Beacon, 1991.

Scheese, Don. *Nature Writing: The Pastoral Impulse in America.* 1995. New York: Routledge, 2002.

Solnit, Rebecca. *Hope in the Dark: Untold Histories, Wild Possibilities.* New York: Nation Books, 2004.

Taylor, Fred. "Loren Eiseley." *American Nature Writers.* Ed. John Elder. New York: Scribner's, 1996. 269–85.

Loren Eiseley

Howard Nemerov

Loren Eiseley
1907–1977

Loren Eiseley and I were what might be thought of as distant friends, a relationship made possible by jet travel, both as to the friendship and as to the distance. We knew one another initially through our books, became acquainted some years ago through some phrases in a book of his which helped me with a poem of mine which he permitted me to dedicate to him on that account, and thereafter corresponded intermittently and had an hour's talk no more than once a year, in his study atop the University of Pennsylvania's Anthropological Museum on Spruce Street, when I was in Philadelphia to talk to a friend's class. So we continued to know each other chiefly through each other's books, and it is about my reading of Loren's books that I will say my small say.

A couple of texts to set in contrast:

> No one objects to the elucidation of scientific principles in clear, unornamental prose. What concerns us is the fact that there exists a new class of highly skilled barbarians – not representing the very great in science – who would confine men entirely to this diet. Once more there is revealed the curious and unappetizing puritanism which attaches itself all too readily to those who, without grace or humor, have found their salvation in "facts."

> Even though they were not great discoverers in the objective sense, one feels at times that the great nature essayists had more individual perception than their scientific contemporaries. Theirs was a different contribution. They opened the minds of men by the sheer power of their thought. The world of nature, once seen through the eyes of genius, is never seen in quite the same manner afterward. A dimension has been added . . .

Both passages come from Loren Eiseley's little book *The Man Who Saw Through Time*, or "Francis Bacon and the Modern Dilemma." I had long thought anyhow that "the man who saw through time" as accurately characterized the author as his subject; Loren, I said, can see through time as easily as I can see across the room. And the two passages, when put side by side, also characterize the position he took up in his several books of meditative and anecdotal and aphoristic essays: a little defensive about his possible or actual, his fancied or real, rejection by "those who have substituted authoritarian science for authoritarian religion;" but committed all the same, with a modest firmness, to the great tradition coming from Sir Thomas Browne (who didn't get elected to the newly founded Royal Society because his prose was insufficiently "unornamental") through such men as Gilbert White and Richard Jeffries and W. H. Hudson down to such exemplars in our own day as Donald Culross Peattie and our late colleague Joseph Wood Krutch.

Before the eighteenth century, when the word "science" was first applied to distinguish its method and objects and attainments from "art," and before the nineteenth century, when Whewell felt a need to distinguish its various practitioners by the one word "scientist," these activities went under the noble if now abandoned names of "natural philosophy" (Newton), "natural knowledge" (charter of the Royal Society), and even "natural theology." As late as my boyhood, children were introduced to science under the appealingly innocent name of "nature study," with its ambience of woodland walks and albums of pressed flowers. Nowadays, the architecture of new science buildings reveals a striking progression; over the past couple of decades, their windows got fewer and narrower, and then were omitted entirely; and the outsider's impression of "science" is made mostly of such devices as oscilloscopes, particle accelerators, computers, and of the activity itself as almost altogether one of counting things and events. No doubt the results are imposing and even astonishing; but one sometimes idly wonders, Where did Nature go?

My own interest in the sciences came on me late, and found me with no preparation beyond the high school in the mathematical languages; so for me the pursuit of even the most modest understanding has seemed a sort of random search in the stacks; and when, some twenty years ago I came across whichever book it was first introduced me to Loren Eiseley's writing, I understood at once that I had made a friend, though the literal sense of that didn't come about till our first meeting a decade later.

I suppose that what we poor literary folk want in the first place is not to be talked to *de haut en bas* by Lord Snow, and certainly not in the second place to be defended by Dr. Leavis, but to be written to on scientific subjects with eloquence, good sense, patience, loneliness rather like our own, and a strong, continuous awareness of the presence of the knower in the known; which is what I have found in a small number of physicists and biologists, and what I derive from Loren, paleontologist, anthropologist, "bone-hunting man" as he described himself, saying "I am a man who has spent a great deal of his life on his knees, though not in prayer," and, finally, poet.

Time for but one illustration of these special qualities of vision and voice that characterized Loren's work, and I choose the one I know best because it worked in me and led on to the poem I spoke of at first that in turn led on to our meeting and too-brief years of friendship.

Years before, I saw a bunch of moths hatched out on an unseasonably mild day on the leading edge of winter. And I wanted them to get over into a poem, only they wouldn't; the best they yielded to my on and off attention over about seven years was on the order of "O you poor buggers, you don't know what you've got into."

But it then chanced that I began reading what became my favorite – a difficult choice – of Loren's essays, a great compound of autobiography and speculation with the mysterious title, as appealing as appalling, of "The Mind as Nature"; where he speaks to start with, as a model for his own early life, of "the dichotomy present in the actual universe, where one finds, behind the ridiculous, wonderful tent-show of woodpeckers, giraffes, and hoptoads, some kind of dark, brooding, but creative void out of which these things emerge – some antimatter universe, some web of dark tensions running beneath and creating the superficial show of form that so delights us." And he goes on to speak, by anecdote and saying, of "that vast sprawling emergent, the universe, and its even more fantastic shadow, life," coming presently to a passage that, among its other effects, put my moths once and for all in place, speaking of "a wonderful analogy . . . between the potential fecundity of life in the universe and those novelties which natural selection in a given era permits to break through the living screen, the biosphere, into reality," and of a hidden world "of possible but nonexistent futures" as "a constant accompaniment, a real but wholly latent twin, of the nature in which we have our being."

The poem thus evoked out of nature and Loren Eiseley's thought was dedicated to him, and may now properly be part of this memorial; it is called "The Rent in the Screen."

> Sweet mildness of the late December day
> Deceives into the world a couple of hundred
> Cinnamon moths, whose cryptic arrow shapes
> Cling sleeping to a southward-facing wall
> All through the golden afternoon, till dusk
> And coming cold arouse them to their flight
> Across the gulf of night and nothingness,
> The falling snow, the fall, the fallen snow,
> World whitened to dark ends. How brief a dream.

Loren was a man of thought and the sayings of thought, and of the loneliness that goes with the life of thought. Of his many wonderful sayings I would end for epitome on one out of his autobiography *All The Strange Hours*, the half despairing, half hoping plea of the ever-only-temporarily defeated Platonist: "In the world there is nothing to explain the world."

Gale E. Christianson

Loren Eiseley in Lincoln:
Two Poems and a Remembrance

I

September 1924 marked the beginning of Loren Eiseley's senior year at Teachers College High School located on the University of Nebraska campus at 14th and S Streets. He had entered the institution the year before, after dropping out of public school, unable to cope with what he termed "the snobbishness" of students who came from Lincoln's upper middle class. They had a future; he did not – or so he had written his young sister-in-law Mamie Eiseley, who was living in Colorado Springs. "Take it from me, boy, I know how that young guy must have felt when he wrote:"

> I never knew sad men who looked
>   with such a wistful eye
> Upon that little tent of blue
> We prisoners call the sky.

"That guy" just happened to be Oscar Wilde, and the adolescent Loren had committed the eccentric Irish dandy's haunting metaphor to memory, as if *The Ballad of Reading Gaol* had been written for him alone.

Conditions in the Eiseley household had deteriorated to the point where the youth had been forced to move into the home of his uncle, the attorney William Buchanan "Buck" Price, and his aunt Grace, sister of Loren's deaf and periodically unbalanced mother Daisy. The youth's social awkwardness was accentuated by the fact that he had reached physical maturity more rapidly than most boys his age, giving him the aspect of one considerably older than his years. Loren was large boned and attained his full stature of nearly six feet within a year or so of entering high school. Though on the lean side, he had a broad chest and powerful, well-developed shoulders and forearms, which made him seem bigger than he was. The chiseled head, too, appears to have been larger than usual, while several of the more distinctive features of the mature man were already in evidence: a broad, prominent nose, down-turned mouth, lantern jaw, and high, granite-like forehead. Loren wore wire-rimmed glasses to compensate for the extreme near-sightedness which otherwise would have rendered him virtually helpless; he oiled his dark hair and combed it straight back in the fashion of the day, taking special care to keep it well trimmed above the ears.

In contrast to Lincoln High School, where the various grades often numbered over 300 students each, those at T.C.H.S. averaged only between 35 and 45 pupils, making it possible to place a premium on individualized instruction. Here the sons and daughters of the professoriat were joined by those born on farms and into Lincoln's rapidly expanding working class. Charles Taylor, the affable principal, personally screened each candidate for admission and was particularly interested in the troubled but promising student, a natural extension of the humanistic credo which made him one of the main supporters of the Salvation Army, as was Loren's Uncle Buck. Taylor regularly took on adolescents who had been categorized by public school officials as incorrigible, academically unfit, or just plain lazy. A few eventually dropped out, but the majority flourished in this protective environment.

While Loren blossomed as a writer of poetry and prose in these pleasant surroundings, he also achieved a degree of social acceptance far beyond anything he had experienced in the past. Not only was he captain of the football team during his senior year, the youth was elected president of his class. Meetings took place in the late afternoon; some students sat at desks while others leaned against the classroom walls. Loren, tall and lanky, would walk self-consciously to the front of the buzzing room and call the meeting to order. No teacher was present, but the banter and repartee ceased immediately. He completed the agenda as quickly as possible, cracking a joke now and then – usually of the self-deprecating variety – but maintaining control throughout. His commanding stature and mature good looks doubtless worked to his advantage, but he was also respected as a student-athlete and well liked by most.

Though his class numbered less than forty, Loren did not chum or mix with everyone. Faye Munden, a close acquaintance and fellow lineman on the football team, characterized him as being "selective" in his choice of friends. He associated mainly with a dozen or so seniors, who commonly referred to themselves as "The Crowd." While the group was made up of both boys and girls, there was relatively little, if any, serious dating because most were planning to continue with their education. On the weekends, they often went to one another's homes for parties. Dancing, which underwent a revolution during the twenties, was among their favorite activities, but Loren, who was somewhat awkward when it came to members of the opposite sex, almost always sat on the sidelines. He much preferred to get out the Ouija board and try his luck with the spirits. A natural with the planchette, he entertained his friends by the hour spelling out messages ostensibly emitted from beyond the pale. One night, while visiting the home of a classmate, he talked the others into accompanying him to the furnace room in the basement, where they seated themselves around a card table. Loren placed his fingers a few inches above the table's surface and instructed the others to do the same, the object being to raise the table off the floor through their collective powers of concentration. The moment of truth was almost at hand when the host's mother, having become suspicious during the prolonged silence, came down the stairs wrapped in a bathrobe, her hair up in braids for the night. Billows of laughter forced her into an unceremonious retreat, and the experiment had to be abandoned, a casualty of the prevailing good humor.

Other than the fact that he was usually short of spending money, Loren's straitened circumstances were not apparent to his friends. His clothes were of the same style and quality worn by others, and he seemed never to have lacked the basic supplies required for the completion of his assignments. The deep strain of melancholy which dominated so much of his writing, both early and late, was also little in evidence. Sixty years later, Loren was still remembered for his wry wit and endearing mischievousness. Georgia Everett, a fellow senior who sat next to him in class, recalled that "he seemed to delight in making people laugh." Though he was neither a prankster nor a disruptive influence, "there was always something going on where Loren was concerned," a view echoed by classmates Leah Dale and Faye Munden. Indeed, he was nicknamed "Bozo" by some of his best friends. Those not a part of his inner circle were encouraged to call him "Larry," Loren's own substitution for his troublesome Christian name, which was sometimes mistaken for that of a girl.

The unofficial title of senior class jester was claimed by John Alfred Cave, Loren's best friend. "Jac," as he was known to everyone, moved to Lincoln with his family early in the fall of 1924, just in time to begin his senior year at Teachers High. Classically tall, dark, and handsome, he sported a carefully groomed mustache, which may have made him the envy of the boys and the object of deep sighing among the girls. A natural athlete, Jac joined Loren and Faye on the football team, shunning the lineman's ignominious lot and opting instead for the spotlight and the glory of the backfield. A great wit and inveterate practical joker, Jac Cave kept the seniors in stitches and his nervous student teachers on the edge of their seats. The mere mention of his name to former classmates still elicits hearty laughter. Helen Hopt, once a member of The Crowd, succinctly but affectionately described Jac as "that rascal."

The second youngest of nine children, Jac was the son of W. Alfred Cave, an Episcopal priest. The Caves were a fun-loving, gregarious clan and treated Loren like another member of the family. Jac's kid sister Alice recalled that Loren was "as dear to us as any brother or son could be." These feelings were reciprocated. Loren warmed to the noisy camaraderie he never knew elsewhere, for Alice thought that in spite of his fey sense of humor, his inquiring mind, and adventurous spirit, Loren was at heart a lonely and rather moody young man.

One of Alice's most vivid memories of her brother's friend was formed when she was eight years old. "I came tearing into the house saying there were two big tarantulas or something hanging on the bushes outside." Jac and Loren followed her out. After a brief look at the creatures, Loren assured her that they were not deadly spiders but harmless polyphemus moths, no doubt blown to earth by high winds the night before. Reflecting on the incident many years later, Alice thought that Loren had displayed considerable knowledge for a youth still in high school. "Perhaps this marked the beginning of my 'crush' [on him] – he seemed to know everything!"

Loren, Jac, and Faye were often seen in the company of another member of the senior class named LeRoy Stholman. A victim of some debilitating childhood disease or injury, LeRoy had partially lost the use of his legs and required crutches to move about. The four thought it would be fun to try their hand at acting, and each succeeded in landing a part in the senior class play, *Kicked Out of College*, a farce by Walter Ben Hare. The plot, such as it was, concerned the marital exploits of Bootles Benbow, a popular college senior who took a different wife in each of the play's three acts. None of the four friends got the lead, which went to their classmate Harold Riggs. Loren played the part of Mr. Gear, owner of the Speed Motor Car Company; Jac was cast as Mr. Sandy McCann, coach of the Dramatics Club; LeRoy played Shorty Long, a member of the Glee Club; while Faye took on the dual roles of Scotch McAllister, "a hard student," and Officer Riley, a cop from the Emerald Isle. The play itself was only mildly amusing, but the cast was royally entertained by the fun-loving quartet. One member of the production recalled that if the public had been allowed to attend rehearsals, the play would still be running.

If Loren did not exactly conform to the description of a melancholy loner during this period of his life, it was nevertheless evident to his more perceptive friends that he was holding something back. Realizing that his family situation was "peculiar," they politely overlooked the fact that he never took his turn by inviting The Crowd to his home on a Saturday night. Faye and Jac got the closest to his relatives of anyone: both met the Prices when Loren was staying at their house on South 23rd Street, but neither ever set foot in the Eiseley home, although Faye harbored vague recollections of once being introduced to Loren's salesman father Clyde, a Willy Loman figure long before Arthur Miller created him. Leah Dale, who sat next to Loren in senior English, remembered that he vowed more than once never to have any children, a vow that was not to be broken.

Some weeks before their graduation in June, "Mr. Gear, Mr. McCann, Shorty Long, and Officer Riley" hatched what was later described as a "harebrained scheme," but which at the time seemed the very quintessence of common sense – if only to the players themselves. They would pool their modest resources and strike out for the West coast, ostensibly in search of summer employment. With their diplomas in hand, the four acquired a 1919 Model T Ford, which was unceremoniously dubbed "Old Purgatory," as if in anticipation of things to come. Other than the extremely modest purchase price, the vehicle's main appeal lay in the fact that its previous owner had altered the front seat so that it would lie flat and make into a "pretty respectable" bed for two. A pup tent, in which Loren and Jac had often slept when camping on the Boy Scout grounds, provided shelter for the others. Before departing Lincoln, the four took special care to stock up on motor oil, since test drives through the nearby countryside revealed that they were burning at least one quart of oil for each five gallons of gasoline. Lacking an oil can of their own, they purchased several gallons from a grocery store, the owner of which obligingly provided them with a container whose bottom and sides were still thickly coated with the residue of its previous contents—wild honey. Thus supplied, the four high-spirited pilgrims started out on what Faye Munden remembered as "a fine June morning."

The first leg of the trip, from Lincoln to Denver, a distance of nearly 500 miles over dirt and gravel roads, took them ten days, but was completed without serious incident. On reaching the Colorado capital, they camped overnight in the yard of Faye's uncle before heading south to the Garden of the Gods, whose multicolored sandstone hills and ridges have been eroded by wind and water into grotesque shapes with fanciful names. They chose to spend the night in the 770-acre park, a decision which nearly resulted in the kind of tragedy long since legendary in the West. Loren woke Jac very quietly the next morning and told him to roll, as carefully as possible, out of the blanket the two were sharing. After they gained their feet, Loren cautiously lifted the cover to reveal a large rattlesnake which had been stretched out between them, basking in the warmth of their bodies.

Following a visit to the Royal Gorge, whose near-vertical walls rise more than 1,000 feet above the Arkansas River, the four resupplied in Canon City before turning southwest. The easy miles were behind them now: ahead lay the San Juan Mountains, part of the Southern Rockies, and primitive Wolf Creek Pass, at an elevation of nearly 11,000 feet. The rock-strewn track – for one could hardly call it a road – snaked its way up the granite escarpment for mile after harrowing mile. According to Faye, it was built for one-way traffic only; "if one met another car and were going up hill it was advisable to find a wide place and put your hub caps right against the rock wall because . . . that fellow . . . needed everything he could muster to get by." They finally made the summit and, thinking the worst was over, began the steep descent in a lighthearted mood. It quickly became clear to all that they had congratulated themselves prematurely. The old Ford's brake bands began to squeal and smoke in protest, forcing whomever was driving to repeatedly shift from forward to reverse as a means of slowing their downward momentum. The transmission bands began to give out and the four breathed a collective sigh of relief when, late in the afternoon, they limped into the little town of Farmington in northern New Mexico.

LeRoy, who had grown up on a farm, served as the mechanic: he spent most of the following day stretched out on his back replacing both the transmission and the brake bands. Their oil reserves were also fast disappearing; and each time the can was tipped a little more honey flowed into Old Purgatory's fouled crankcase. To make matters worse, the tires had taken such a beating that flats had become a matter of routine, as had the scraped and bleeding knuckles acquired in the process of repairing them.

For some unexplained reason, no one had thought to bring along a jack, and there was too little money left to justify the purchase of such a "frill." A simple but effective system was devised: "Loren . . . would back up against the wheel, take hold of the wooden spokes, and lift that quarter of the car high enough so we could [put a] block under the axle and change the tire." The spare soon went, forcing them to buy another. Even gasoline proved a problem, because the petroleum industry was only just beginning to ship refined fuel into this still remote part of the Southwest.

The undaunted quartet pushed on through Sante Fe and connected with Route 66, which was then known as "Old Trails Highway." After passing through Gallup and crossing into Arizona, they decided to visit the Petrified Forest. Turning south off the main road, they soon came to the Rio Puerco. There was no bridge in sight. It looked like other cars had simply driven on through, and they followed suit, luckily avoiding the hidden rocks and sinks of quicksand. They spent most of the day exploring the six separate "forests," with great logs of agate and jasper lying on the ground surrounded by the varied colors of endless fragments and small chips. Dating from the Triassic period some two hundred million years ago, these "stone trees" had died from natural causes, such as fire, insect damage, and fungus rot. After toppling, they were deeply buried in sand and silica rich in volcanic ash. Minerals carried into the cells by ground water slowly turned wood to stone. Finally, as the surrounding material was eroded away, the petrified trees were exposed to the surface once again. Prehistoric Indians had once lived among these stone giants; the remains of their dwellings and haunting petroglyphs were clearly in evidence. The four stayed late into the afternoon, reluctantly fording the Rio Puerco a second time and heading west into a brilliant Arizona sunset. Awed by what they had observed, no one said very much – least of all Loren, who had seemingly withdrawn into a world of his own.

The youths got as far as Peach Springs on the Hualapai Indian Reservation before the honey finally caught up with them. They camped in a park by the railroad tracks, where LeRoy proceeded to undertake a major overhaul of the motor. This wasn't their only problem: they were flat broke. Faye, whose father and grandfather were railroad men, persuaded the local Sante Fe agent to let him catch the "California Limited" to Kingman free of charge. "I got the agent at Kingman to wire my Dad (free again). Dad was so pleased to know I was still alive he sent money and a day or so later we had Old Purgatory going with fresh parts and clear oil."

They had driven only about 100 miles when disaster struck in the Mohave Mountains just outside of Oatman, near the California line. No effort was made to control the steepness of grades in those days, and the old problem of worn brake bands suddenly cropped up again. Faye was behind the wheel and heading down the side of a mountain. "At first I was able to handle the grade pretty well." But in the distance, the road made a sharp turn to the right which required that the car be slowed almost to a stop. Faye stepped on the brakes; nothing happened. Unable to negotiate the hairpin, he had two choices: either to drive straight ahead into a towering wall of rock, or to angle left on a strip of upgrade littered with massive boulders:

> By that time the boys could see what was going to take place and they started to unload. Loren jumped off the right side and slid on his stomach with his hands ahead of him. He skinned his palms and hands pretty badly. Jac got off without injury as I recall but LeRoy [who was crippled] rode with me as I turned to the left up the grade and into the boulders. Well, after we negotiated three or four of these 4 foot boulders we came to a halt wedged between 2 good sized ones. All tires blown out and the car a complete wreck. LeRoy and I were scared but all in one piece.

The four walked the remaining distance into Oatman and stopped at the town's only garage. They asked the owner what he would give them for Old Purgatory, sight unseen. "I guess because he felt sorry for us he offered $5.00 cash. We took it and got something to eat."

Later that same day, they hitched a ride into Needles, California with two miners who were driving an old Jordan touring car. Only after they set out from Oatman did it become apparent to the youths that their benefactors had been drinking. The boys were taken for what Faye described as "a merry ride through the winding dirt roads of the desert." They had gone about eight or ten miles when the right rear of the car started to vibrate and rattle. Someone pointed out the window and yelled, "There goes a tire past us!" The tipsy driver somehow managed to stop without doing harm to either the car or its occupants. The tire was retrieved and mounted with lugs borrowed from the three other wheels. It was mid-afternoon and steaming hot when the six finally rolled into Needles. The boys wasted no time heading for cover: "We found shade down by the Harvey House [Hotel] and the temperature was away over 115°."

Towards evening, they hopped a freight and rode to the first water tank, where the brakeman put them off. When the next westbound train came through, they got down inside the compartment of a refrigerator car which still had a little ice in it. Although it was dark as "billy hell," the welcome break from the stifling desert heat made the temporary sacrifice of sight well worthwhile. Several hours later found them "yarded" in the isolated town of Barstow, some eighty miles north of San Bernardino. When it became evident that the train was not going anywhere, they began to climb out of their hiding place. As Faye was about to emerge through the roof the hatch door came down on his head, leaving a nasty gash above the scalpline. It bled profusely, forcing him to find a water hydrant so that he could wash up.

It was apparent to everyone that LeRoy's disability was incompatible with jumping freights. They wired home for more money and, when it arrived, reluctantly put their boon companion and star mechanic on an eastbound passenger train. Loren, Faye, and Jac made it the rest of the way to Los Angeles, where one of them rented a single room for $4.00 a week, concealing from the landlady the fact that all three were living together. Money remained tight, and the trio took whatever work they could find, however menial. Faye remembered that he and Loren were employed by a contractor to dig ditches, while Jac, a more genteel sort, clerked in stores. Still, they could not make ends meet, and the telegraph lines stretching back to Lincoln a half a continent away were kept humming with further requests for financial aid. Faye decided to call it quits after a short while and used the last of his money from home to purchase a ticket to McCook, Nebraska, to which his parents had moved immediately following his graduation. Though he had no inkling of it at the time, Faye, who eventually became an engineer for the Santa Fe Railroad, would never set eyes on his high school friends again.

Little is known of Loren's and Jac's movements during what remained of the summer, except that they stayed on in California awhile longer. According to Alice Cave, they may have moved down the coast to Long Beach, so that Loren could be nearer to the ocean whose shores he had envisioned walking ever since he first tiptoed into his Aunt Grace's bedroom and held her mysterious iridescent seashell against his ear. All that is known for certain is that he persuaded Jac to undertake one final journey before heading back home. They walked and hitchhiked most of the day to reach the famous Mount Wilson Observatory northeast of Los

Angeles. According to the guidebooks Loren had read, the public was allowed to visit the complex on certain nights and look upon some remote object through the giant reflecting telescope. They arrived, exhausted but hopeful, and not a little naive. The guard eyed them and their disheveled clothing with "sullen distaste." Loren became keenly aware of what Bret Harte meant when he wrote of the defective moral quality of being a stranger. "We thought, though we were poor, that we would be welcome upon the mountain because of our desire to learn. There were reputed to dwell in the observatory men of wisdom who we hoped would receive us kindly since we, too, wished to gaze upon the wonders of outer space."

The two waited through the night, as busload after busload of tourists arrived from the better hotels in the valley to be greeted by uniformed guides. The mountain air was freezing cold, but it was equally plain that the freelancers were not welcome in the inn, which also catered to the tourists. They purchased some chocolate and waited outside until the glint of the rising sun pierced the eastern horizon, confirming what both already knew. They would not be allowed to see the men of wisdom after all. The two looked at each other and then, wearily, without saying a word, turned and began their long descent through the early morning dark.

Throughout Loren's life women were more closely drawn to him than were men. This surely had much to do with his physical attractiveness and deep contradictions of character – the scarcely disguised vulnerability and constant need for emotional support on the one hand, the natural charm and seeming worldliness on the other. A disciple of Narcissus, he returned their attentions in kind, and liked nothing so much as to be surrounded by what might be called a protective nest of femininity. When his spirits flagged, he drove over to Nebraska City on the Missouri to be alternately teased and praised by a group of young women who laughingly referred to themselves as "Loren's Harem."

The group was composed of sisters, girl friends, and in-laws of Jac Cave, whose father had been chosen to head the local Episcopal congregation in 1926. Loren, now a university undergraduate, usually visited on weekends, and loved to pose for photographs with them. On one such occasion he donned a long apron for comic effect and shaped his dark hair into a formidable spit curl; the harem then lined up four on either side. During another of his visits in 1932, Jac's sister Alice, now a high school sophomore, asked him to write something in her autograph book, a popular fad of the times. She expected the usual nonsensical jingle such as, "Way back here where no one will look, I'll sign my name in your autograph book." The adolescent was amazed on returning home for lunch to find Loren wandering about the house in the throes of composition, "a far-away look in his eyes and unintelligible murmurings on his lips." The finished poem, "For Alice," was never published and remained unknown outside the Cave family for over fifty years:

> You said you liked John Keats. Take this thought home.
> Into your heart – who have some years to live:
> Beauty has fox feet and she always runs
> Down hidden roads – a harried fugitive.
>
> Beauty has fox-feet and a rippling coat
> Goes with her on her way into the night.
> A quick, wild muzzle fallen in the brake
> Or hawk's wing broken on some bitter height
>
> Betrays her passing. Let no winter shut
> Your mind against her. Let no pain revise
> Her printed footsteps on a bed of fern,
> Or wind on air – or tempt you to be wise –
>
> Oh, soon enough she seeks another place –
> Goes southward with the wild birds, Alice, goes with wings,
> Like childhood, like the swift green leaf that's blown
> Into the dust that never cries or sings.

Alice caught her breath and blushed: "That this handsome, sensitive, adored friend of my beloved brother should take so much time and care to write so understandingly for Jac's kid sister . . . seemed to me the epitome of everything that was kind and good and worthy to be adored. Which, of course, it was." Alice was crushed when he brought the pretty Mabel Langdon along on one of his visits. The older woman, who would one day become Loren's wife, didn't stand a chance. "What little attention I paid to her was more resentful than anything else. I remember her only as quiet, very reserved and, to my young mind, colorless when compared to our boisterous clan."

Jac lost contact with his friend a year later. The minister's son drifted through the long Depression, taking whatever job offered itself – insurance salesman, truck driver, hotel desk clerk. Before leaving Nebraska for good, he dreamt up the publicity stunt of floating down the Missouri, from Omaha to St. Louis, in an inner tube. While the gimmick made him no money, he was given a full-page spread, complete with pictures, in the Omaha paper. Loren almost certainly read the story and doubtless smiled to himself: the senior class jester was still living up to his reputation. Then the handsome young man drifted quietly down memory's river, not to be heard from again until after Loren published his autobiography more than half a lifetime later.

II

Wilbur Gaffney, an elfinlike young man of dark features, dancing gray eyes, and puckish wit, joined the editorial staff of the University of Nebraska's literary quarterly, the *Prairie Schooner*, in 1927, at the same time as Loren. A poet in spirit but a satirist at heart, "Bill" had contributed a few irreverent pieces to the campus humor magazine *Awgwan*, a corruption of the then popular catch phrase, "Aw, go on!" Though he had once entertained hopes of joining the *Awgwan's* staff and becoming a regular contributor, the Lincoln northsider from the blue collar class soon realized that growing up in a cold water flat above a meat market counted for little when one was pitted against "the self-perpetuating monopoly of the fraternity set." Undaunted, he cheerfully hitched his literary star to *Prairie Schooner*, where Professor of English Lowry Wimberly and his student editors cared nothing about one's social and economic background. It was here that he met Loren, "who was not a proper southsider but a working man's son. That gave us a certain amount of kinship." The much taller and more muscular Eiseley also traveled by boxcar, something Bill had always wanted to do but was afraid to attempt. Finally, both loved to take long walks across the salt flats and gently undulating countryside surrounding their native Lincoln.

Exactly when Loren and Bill took their first hike together is no longer remembered, but it could not have been long after the two became acquainted. Their most ambitious excursions, which were reserved for Saturdays, took them southwest of the city to the bucolic village of Denton, a distance of some ten miles as the crow flies. They got the idea of going in this direction from English professors Kenneth Forward and Sherlock Bronson Gass. However, unlike their mentors, the youths did not have the money to catch the 5:00 p.m. train back to Lincoln, and so retraced their steps for a total of twenty or more miles. Their standard fare on these expeditions was a bar of German chocolate – nourishing, but not

so sweet as to arouse thirst. The high hills gave them an unob-
structed view of the city skyline, which was dominated by the new
state capitol. Near Denton they crossed three or four miles of un-
fenced land, which the two fancied was perhaps the closet thing
to moors they might ever see. The English and Irish poets came
automatically to mind; they quoted stanzas by Tennyson, Coler-
idge, Wordsworth, and Yeats from memory.

On other occasions the two ventured north of town, via 14th
Street, to the old military academy on Belmont Hill, then back along
the Northwestern tracks bordered on either side by deep blue ponds
shimmering in the afternoon sun and ringed with gleaming de-
posits of white salt. Bill remembered the exact spot on Belmont
Hill where they once paused while Loren introduced him to Chesterton's *The Ballad of the White Horse*, quoting a haunting stanza
about the invasion of England by the pagan Norsemen with their
Christless chivalry:

> And men brake out of the northern lands,
>   Enormous lands alone,
> Where a spell is laid upon life and lust
>   And the rain is changed to a silver dust
> And the sea to a great green stone.

Young Gaffney was so moved that he later committed much of
Chesterton to memory, including all of the beautiful "Lepanto,"
which he quoted to an appreciative Loren in return.

It was also atop Belmont Hill one day that the two experienced
an interesting coincidence Bill long remembered. In 1924 the *Lincoln
Sunday Journal* printed the three closing stanzas of Franklin Mc-
Duffee's poem "Michelangelo," which won Oxford University's
Newdigate Prize. Bill had lost his clipping, but he had memorized
the stanzas and he recited them to Loren in hopes that his friend
might be able to identify the author and title. It so happened that
Loren, too, had saved the clipping and told Bill what he wanted
to know. Indeed, he had committed the same lines to memory and
joined his friend in their recitation. He then said, "I've done you
a literary favor; maybe you can do me one. Do you know a poem
that appeared in the *Sunday Journal*: 'How the pale lamps burn in
the valley; the dim dust creeps . . .' " Bill laughed heartily and
replied, "I can help you with that; I wrote it."

On their only long walk east of town they fought their way
through a thorny hedge of Osage orange, half a mile long, and
trod across barren farmland in an enveloping early December fog.

They finally reached the banks of Stevens Creek, already frozen solid. Chilled to the bone, they built a fire on the thick ice and sat for a time telling stories and quoting their favorite poems. Shrouded in fog, all landmarks obliterated except for a silver sliver of the inanimate stream, Loren and Bill experienced "the delicious sensation of being absolutely lost" when the time came to go.

While they agreed on many things, the two friends sometimes differed in their opinions about the new generation of poets. Whereas Bill was tepid when it came to the works of Robinson Jeffers, Loren was enthusiastic about the alienated master of Tor House – the personification of twentieth-century man who knew too much. Jeffers had moved so far west that he came to the end of the land. At Carmel, California, on the Monterey Peninsula, he built a granite house and observation tower with this own hands. Here he worked in splendid, albeit tortured, isolation, the mountains and sea his backdrop. Humanity, Jeffers asserted in his poem "Contrasts," is inferior to animals and natural objects: "There is not one memorable person to stand with the trees, one life with the mountains." The human race has become introverted and egocentric, incapable of recognizing the transcendent significance of other creatures and things in the universe. A superior reality exists behind appearances, a reality that can be discovered, though only by suffering great mental anguish.

Loren was especially drawn to Jeffers's short poems, but he also read the longer narrative works, which deal with murder, incest, and other kinds of violence. Bill remembered his reciting "To the Stone-Cutters," men who fight time with marble, "foredefeated challengers of oblivion."

> The square-limbed Roman letters
> Scale in the thaws, wear in the rain. The poet as well
> Builds his monument mockingly;
> For man will be blotted out, the blithe earth die, the brave
>   sun
> Die blind and blacken to the heart . . .

Sir Thomas Browne, the seventeenth-century author and physician, had communicated a similar message in his *Hydriotaphia* or *Urne Buriall*, which Loren and Bill both read and admired in a course on the great English essayists. This solemn reflection on death was inspired by the discovery at Norfolk of several Saxon burial urns, which were mistaken for Roman artifacts. Chapter five made a particularly vivid impression on the young men. "Circles and right lines limit and close all bodies, and the mortall right-lined circle must conclude and shut up all. There is no antidote against the *Opium* of time . . . which maketh Pyramids pillars of snow, and all that's past a moment."

So far as Bill could tell, Loren's interest in the darker side of human existence did not make him particularly morose. In terms of psychological balance he seemed "fairly normal," though more stoic than emotional. They were both to know young men who took their own lives, but Loren shared none of their characteristics. "As to his pessimism, he had the same sense as Jeffers had [that] the world is gradually going down hill and we will all wind up in the grave, forgotten except for a few inscriptions." Still, this ingrained sense of resignation kept neither Loren nor his poetry-writing hero from facing the new day.

All was not gloom and doom on the "moors" and salt flats of eastern Nebraska. Loren seemed especially fond of the Irish poet and novelist James Stephens. Under five feet tall, balding, with an elongated face, crackling brown eyes, and dark skin, Stephens possessed a sardonic wit which cast him in the perfect image of a leprechaun. Loren memorized several of Stephens's poems and quoted frequently from one of the Irishman's shorter works, "A Glass of Beer." It concerns a perpetually inebriated Dubliner whose incessant pleas for a pint or two of free stout receive no sympathy from the neighborhood barmaid. Threatened with a permanent loss of his self-esteem at the hands of this woman, or, as he refers to her, "that parboiled ape," the old tippler has only one recourse – the Irishman's curse:

> May she marry a ghost and bear him a
> kitten, and may
> The High King of Glory permit her to
> get the mange.

They were also interested in certain of the more popular humorists. Their favorites were A. A. Milne, the creator of the immortal Christopher Robin and Winnie-the-Pooh, and Don Marquis, the protégé of Joel Chandler Harris, author of the *Uncle Remus* stories. Loren and Bill both followed the syndicated adventures of Marquis's most unlikely characters, Mehitabel, the alley cat who believed her soul was formerly incarnated in the body of Cleopatra, and Archy, the cockroach poet and story teller who typed all of his manuscripts without punctuation and in the lower case because, according to a well-known law of physics, no cockroach could hop on two keys at the same time. Archy's poem, "pete the parrot and shakespeare," tickled both young men. The parrot, according to Archy, had once belonged "to the fellow that ran the mermaid tavern in london." The venerable old bird knew "shake-speare," or "bill" as he was wont to call him. As is often the case in history, things are never quite as they seem:

> says bill what the hell
> is money what i want is to be
> a poet not a business man
> these damned cheap shows
> i turn out to keep the
> theatre running break my heart
> slap stick comedies and
> blood and thunder tragedies
> melodrammas . . .

The two sometimes made up some rather risqué limericks, which they recited back and forth. These hardly approached the vulgar, however. Rudolph Umland, a keen observer and gifted writer who spent considerable time with Loren during the thirties, observed: "I never heard him use a dirty four-letter word but he occasionally said goddam and called certain individuals he didn't like bastards."

Loren and Bill always carried notebooks in their shirt pockets just in case their muse should chance to pay a call. At times these blank pages were put to less exalted use. They would sit down on the banks of a pond or a stream and fashion little paper boats, as they had done when they were boys. The fragile craft were then placed on the water and watched until they capsized or disap-peared beneath the rippled surface. The two especially loved the north side of Capital Beach Lake, located on a salt flat. They walked the gently sweeping arc of rimed shore, occasionally bumping shoulders, Loren pointing out flora and fauna that Bill "wouldn't have begun to recognize." The prevailing summer wind blew salt spray into their deeply tanned faces, while an inland species of gull darted and cried in the distance. For a brief while it seemed to them that Nebraska was partly covered again by the great sea that had once washed over everything from central Mexico to the south-ern borders of Canada. Short man and tall, light of foot and heavy, they turned homeward reluctantly, Tennyson's lines of poetry, which Bill was fond of quoting on occasions such as this, wafting on the freshening breeze:

> In the afternoon they came into a land
> in which it seemed always afternoon.

Kenneth Forward, a diminutive, soft-spoken professor of English, used to recall a walk he took with Loren in 1932. Eiseley was then a student in Forward's class titled The Nineteenth Century Essay, and he had turned in a paper describing the houses in which he had lived while growing up on Lincoln's south side. All three homes were within a few blocks of one another; each held painfully vivid memories of his mother's presence. Loren's essay so touched Forward that he talked his student into taking him on a tour of the neighborhood. Five or six years later, Forward retraced the route he had taken with Loren for the benefit of Rudolph Umland. When they paused in front of the Price's house the curtains parted, as if on cue, and a gray face peered out. The embarrassed pair quickly turned away and resumed their walk. "The stone-deaf woman and her sister were still living in that house."

While Loren's reputation as a writer was largely that of a poet, time would one day reveal that verse represented his passport to prose. Forward's course, which provided the disciplined setting in which to experiment with the essay form, dealt exclusively with the English masters: Hazlitt, Macaulay, Lamb, DeQuincey, Carlyle, Ruskin, Huxley, Newman, and Stevenson. His lectures covered the biography of each of the essayists, their respective differences of style, and an analysis of their finest literary endeavors. Students were required to compose at least five essays a semester, adopting the prose style of the various figures assigned to them. Though Forward was a precisionist when it came to language and punctuation, he was certainly no pedant. Bill Gaffney, who took the course a year before Loren, remembered that Forward "didn't insist on content, so long as what we wrote was literate and showed that we had some understanding." On one occasion Bill could think of little to say about Robert Louis Stevenson, even though he greatly admired his genius. "I evolved a page or two of whimsy in which I had more footnotes than text; I got full credit, plus a whimsical commentary."

It so happened that Loren took a somewhat greater risk, partly because of Forward's reputation as an easygoing sort, and partly because the young professor sometimes walked to Denton with Gaffney and Eiseley on Saturday mornings. Loren was told to write an essay in the manner of Walter Savage Landor, a little-remembered poet, literary critic, and prose writer, known for the severity and intellectual aloofness of his style. When the time came to turn in his assignment, Loren submitted not an essay but a long forgotten poem titled "Lizard's Eye."

The dust devils spinning in the alkaline basins of drouth
    north of Mohave
and the desert Joshua's curious pleasure in thorns
are a pillow of comfort after a wasted lifetime.

I could come back here and be content with my friends the
    lizards
to burn in the lime-glare and have fierce wild thoughts like
    the animals,
nor much remember
your face, nor the tossed black head of you,
nor security's penny, nor the lie which has left
you beautiful, me, ash and blown fire,
Not the stained agony, pain, nor the casual whore at the
    street corner,
nor the whirled emptiness of death
looked into
roils peace when one's grown accustomed to burning
and seen life with the lizard's eye
not blinded by sun.

He appended the following note, dated January 28, 1932, to the bottom of the page: "I had something crawling up and down in me last night – and this is the result. I thought you might not mind it in place of the Landor essay." Instead of failing the paper or making Loren do it over, Forward displayed his usual equanimity. He gave Loren an A, the same mark he received for both semesters of the course, a gesture the young writer never forgot.

Of the essayists whose literary style Eiseley was required to imitate, none seems to have left a more lasting impression on him than Thomas H. Huxley, whose advocacy of evolutionary theory earned the memorable Victorian the nickname of "Darwin's Bulldog." Indeed, Forward delivered one of his most penetrating lectures on the virtues of his favorite Huxley essay, "On a Piece of Chalk." Bill Gaffney was so impressed that the future professor of English found himself lecturing enthusiastically about it forty years later, while Rudolph Umland, who also took Forward's course, observed: "I imagine Loren never shook 'On a Piece of Chalk' out of his head. I know I haven't." On another occasion Umland wrote that Huxley's essay "must have sent [Loren] musing and pondering under the stars . . . Thirty years later [he] became the Huxley of his time but a Huxley with more poetry in his soul."

"On a Piece of Chalk" was first given as a lecture to the working men of Norwich in 1868. "Huxley," in Eiseley's own words,

"understood instinctively how to take simple things – in this case
. . . a piece of chalk in a carpenter's pocket – and to proceed from
the known, by some magical doorway of his own divising, back
into the mist of long-vanished geological eras." Huxley told his
rapt audience that when placed under a microscope, a thin slice of
chalk presents a totally different appearance. One sees that it is
composed of innumerable minute bodies less than a hundredth of
an inch in diameter, each having a well-defined shape and structure. These granules, or *Globigerina*, were once living organisms
which drifted, like microscopic rain, from the surface of the world's
great oceans to the dark, muddy bottom below. Thus did the layers
accumulate, time out of mind, while great geological catastrophes
intervened between one epoch and another. In the days before the
chalk broke through to form the familiar cliffs of Albion, "the beasts
of the field were not our beasts of the field, nor the fowls of the
air such as those which the eye of man has seen flying." As some
species vanished, others took their place. "The longest line of human ancestry must hide its diminished head before the pedigree
of . . . insignificant shellfish."

It was writing such as this that pointed Loren in the direction of
the popular scientific essay, that helped him to shape his own
grand metaphor of limitless change in limitless time. The grind and
crush of glaciers would one day capture the poet's imagination, as
he raised his sleepless head from the pillow in the small hours to
listen for the sound of night frost splitting stone in some nameless
valley far, far away.

Erleen J. Christensen

Loren Eiseley, Student of Time

In 1928, Loren Eiseley made his literary debut in *Prairie Schooner* with "Spiders," a poem in which he says, ". . . time is a spider, / the world is a fly . . ." Eiseley changed his metaphors as he matured, but his preoccupation with the philosophical meaning of time continued, and he explained why in an address first given in 1954, at the height of his career:

> Yet it is not the fascination of time that propels me. The ache and nostalgia of one vanishing lifetime are a sufficient burden without multiplying that pain a thousand fold. Rather, it is a pain which has to be borne incidentally in the search for the meaning of life, because life in some mysterious manner is involved with time. It is not for the pleasure of seeing men as beasts, that one makes the journeys backward, nor is it the naturalists' hunger for the uncorrupted world before the cities. It is not in malice, not in contempt that the journey is made until the lights of the present grow dim and far as galaxies the outer edge of space. One makes it instead, in the simple effort to understand the nature of life, to go rung by rung down the full length of that terrible ladder until one stands in the brewing vats where the thing was made amidst strange acids and delicate crystals. (Unpublished Phi Beta Kappa Address delivered Jan. 13, 1954, Papers of Loren Eiseley. Philadelphia: University of Pennsylvania Archives.)

Fourteen years later, in the last book published in his lifetime, the autobiographical *All the Strange Hours*, Eiseley has moved yet farther back, to "the country of vertical time" governed by the stern warning, *"Behind nothing, before nothing"* (*Hours* 263).

In *All the Strange Hours*, Eiseley deals with time on several historical levels. The chapters follow Eiseley's life reasonably systematically from the time of his father's death to his own death, or rather his accounts of experiencing his own death in a dream. Yet the book actually opens by describing a day in 1974 when Eiseley first felt old; the actual time during which the narrative was put together, 1974 and 1975, provided the basic historical time frame for the book. Each of the three sections of the book begins with a death and ends with a stepping out of historical time.

In trying to explain how the brain, especially the brain of a writer, deals with time and memories, Eiseley says:

> In all the questioning about what makes a writer, and especially perhaps the personal essayist, I have seen little reference to this fact; namely, that the brain has become a kind of unseen artist's loft. There are pictures that hang askew, pictures with outlines barely chalked in, pictures torn, pictures the artist has striven unsuccessfully to erase, pictures that only emerge and glow in a certain light. They have all been teleported, stolen as it were, out of time. They represent no longer the sequential flow of ordinary memory. They can be pulled about on easels, examined within the mind itself. The act is not one of total recall like that of the professional mnemonist. Rather it is the use of things extracted from their context in such a way that they have become the unique possession of a single life. The writer sees back to these transports alone, bare, perhaps few in number, but endowed with a symbolic life. He cannot obliterate them. He can only drag them about, magnify or reduce them as his artistic sense dictates, or juxtapose them in order to enhance a pattern. One thing he cannot do. He cannot destroy what will not be destroyed; he cannot determine in advance what will enter his mind. (*Hours* 151)

In his autobiography, Eiseley skipped wildly through time and space. He wrote in loving detail of things that happened only in dreams and said nothing of key milestones in his professional and personal life. But he built into the warp and woof of his autobiography the tension between conventional historical time and the older cyclical time of primitive peoples and our own cluttered memories. And behind it all, all that panorama of life and memory, he saw the procession of the time-bearing gods, "behind nothing / before nothing / worship it, the zero." Eiseley quotes those lines of poetry from a mysterious "Someone" who will later prove to be Eiseley himself. The full poem appears in *Another Kind of Autumn*, the book of poetry published just after his death.

The idea of worshipping the zero was a brilliant attempt to understand the nature of time. It tries to explain the puzzle of how something came out of nothing, how order came out of chaos, how life came from non-life, by understanding the concept of the zero. Zero allows man to count in the millions and billions and give physical representation to nothing. Zero is nothing as an entity that can be perceived and manipulated, an invisible placeholder that makes all the difference in counting and in perceiving the immensity of time and space. When we combine zeroes with the integers, those numbers we can actually see as we count on our fingers, we are able to measure time, space, and volume in amounts too great for either our eyes or our limbs to fathom.

In "The Maya," the poem about a people who worshipped the zero as a god, Eiseley wrote of his admiration for those people who invented the zero, and said:

> They worshipped time and zeros
>   constellations crawling
> endlessly above the rain forest,
>
>   marked a leaf's descent,
>   a flower closing,
>   but knew
>
> time was the god
>   added another zero and pressed back
> beyond their own beginnings
>   erected the wrought stone
>
> to mark a million years before they came
>   a million after (*Autumn* 23–24)

The Maya Eiseley eulogized in that poem had died out many years before our day, but they lived their own little space in time, felt joy, loved beauty, built cities, buried their dead with pomp while, unlike us, they had a sense of the immensity of time, of how small they were in relation to that which they worshipped. Eiseley revered a people who could see the immensity of the universe with such scientific rationality that they invented and used the zero long before western civilization had advanced that far, yet he also revered the Maya for their ability to embrace such a concept with a religious awe that worshipped the immensity of time, marking the place of constellations a million years before and after their own time with their stelae to the zero:

> . . . What other people
> have had the strength for this
>   and quietly
>
> let their hands fall
>   obliterating the secret of the markers
>   erasing the constellations
>   before their disappearance –
> a blackboard exercise a god might have envied.
> (*Autumn* 238–40)

Eiseley's career as a writer was, in large measure, an attempt to explain his expanding concepts of time, concepts that were not intellectual exercises but a form of religious belief. From his very first poem, "Spiders," Eiseley was trying to deal with death and loss by seeing the larger perspective, feeling part of some larger time span than a human lifetime. The sequence moved fairly quickly out of the historical time of written record and the civilizations that have given us the word *history*. Evolution allowed Eiseley to look at the larger time span in which animals are our ancestors and brothers, yet in the end Eiseley knew that evolutionary time, like historical time, was nonrecurring, that the genetic dice that cast up a particular individual or species could not be rethrown. By the time he published *The Invisible Pyramid* in 1970, he preferred the cosmic time of the stars and the comets, the cyclical time concept of early tribes to historical and evolutionary time.

Eiseley always saw himself as moving backward as he developed understanding. He saw the tree of life extending backward from his students and himself; he followed Haley's Comet backward to the time when men had not yet crossed the bridge to become human; and in his autobiography he tries to imagine the chaos, the antimatter, the zero out of which we came, into which we travel. The simple concept of the zero as god is important, but so is the journey backward through the human mind. Memories interrupt Eiseley's rational account of his life history; causal connections bring up still other old memories; and sometimes tiny unrelated fragments, dreams and other experiences not part of his conscious life, move him farther backward, or downward into the lower levels of human consciousness.

In the opening portions of *All the Strange Hours*, Eiseley's reference to the zero as god is rational, a quotation from a poem, a part of a speech on the development of human civilization. In the closing chapter, it is not an abstract concept but the destination toward which his own being journeyed in dreams, would journey after death.

In the recurring dream of the last chapter, Eiseley is moving back. He says, "I did not care to be a man, only a being." Going on, he adds that he will "dream again, but further, further back." As he moves backward through the snow of time, he will be behind the historical time in which men with rifles shoot down convicts; he will even be behind the evolutionary time when men and animals became the same (*Hours* 265–66).

Each of the three sections of the book displays this pattern of transcending nonrecurring time. In "Days of a Doubter," the young man, at the beginning, rides the fast freights to get back to his dying father, yet he only arrives in time to see the father acknowledge the presence of a half-brother and leave Loren standing, unrecognized.

At the end of the "Days of the Doubter" section, the young man enjoys "The Most Perfect Day in the World," calling it that because it is "the day time stopped" (*Hours* 66). Stepping out of time and its concomitant pain was the perfection for which Eiseley strove. On the surface the most perfect day was unspectacular, merely a drowsy autumn day when four young drifters lounged on a loading platform in Kansas, drinking grape soda. But Eiseley imbues the experience with a symbolism that made that day a microcosm of human history. He sees the four young men as four races of men – the Indian, the Egyptian, the Greek, and himself, a person out of "the unformed malleable present" (*Hours* 63). The four had no past and no future there on the platform; even the present did not intrude in the form of passing trains or harassing officials. The four simply *were*. They were not going anywhere – and Eiseley as an old man could remember no other day in his life so close to perfection.

The section called "The Days of a Thinker" moves in a similar way from an extreme awareness of historical time to a total indifference to it. This section opens with a chapter called "The Laughing Puppet," an account of a time when Eiseley passed out on the steps of his boarding house with a serious illness; it ends with an account of a retarded man who is bewildered that his childhood playmates no longer come to play with him. That retarded man in the chapter called "The Palmist" cannot conceive of time. To him, thirty years have not passed. He would never recognize the adults his childhood playmates have become; he simply remembers them as they were thirty years ago and wonders that they no longer come to play. After telling of the retarded man with no sense of time, Eiseley continues telling of a fortune teller who warns, "you will die by water." Disjointed and associational, the last portion of "The Palmist" moves from topic to topic; but overall it is about memory, about the brain "talking to itself, carrying on some vast dialogue I was incapable of deciphering." Toward the end of the chapter, Eiseley describes a time in childhood when he found some dice in an abandoned house and played with them. At the end of the chapter, he describes going into a bar and drinking, with the toast, "To whatever I won with the dice in childhood. . . . And to the last cast" (*Hours* 213).

The dice game of childhood, of course, connects intimately with the idea of worshipping the zero. The child played against no visible adversary, following no discernible rules, yet he felt the compulsion that he must *play*, that the game he played was not an illusion but a contest that mattered. As Eiseley became older, the Other Player against whom he played such a game grew grimmer and more real. Eiseley felt a presence, at times saw it – yet that opponent, too, was nothing, a zero. As a young man working in a hatchery, Eiseley had felt a vague presence in another aisle, and while apprehensively stalking this nothing had discovered a fire which could easily have killed him. At about the same time in his life, Eiseley had a vivid dream of fighting desperately with an adversary, and finding himself, at dream's end, throttling a creature which had shriveled into an evilly grinning puppet. Still later, a figure on the other side of the campfire from him "raised his cowled head at my uneasy stare. Beneath the hood there was no face, nothing, merely a chill like the void" (*Hours* 220).

In the final chapter of *All the Strange Hours*, the Other Player comes in a dream, speaking to Eiseley of the game he played in life, telling him that he is in "the country of vertical time," where one does not go back. Yet even the face of the Other Player fades as Eiseley moves back to a time when "The rifles will be silenced, the dice at last unshaken" (*Hours* 266). The Other Player is nothing, both in the colloquial sense that he does not matter and in the mathematical sense that zero represents. The thing we fight and fear and the thing we can worship because of its power to represent infinite space and time are, after all, the same.

Eiseley closes his autobiography with a certain sense of victory in the face of death. Beside him is the dog, Wolf, an animal reminiscent of the animal familiar or guide of the Native American cultures. They outwit the Other Player and get past the posse waiting to gun them down; they reach the place where "the carefully drawn human lines would be erased between us." Even "Coyote the trickster, who is unscrupulous and wins at gambling" would be there in the deep snow beyond human time (*Hours* 266).

In the very first chapter of *All the Strange Hours*, Eiseley introduced the "rat that danced," the animal that stole the show with his antics on stage when Eiseley gave a lecture on the four great civilizations – in the year when Eiseley first felt old. Eiseley's dualism was neither the Christian battle between good and evil nor the Oriental yin and yang. After the lecture, as an old friend told Eiseley of the dancing rat, they spoke of the trickster, and Eiseley says,

> Wasn't it I who had once written that there was a trickster in every culture who humbles what are supposed to be our greatest moments? The trickster who reduces pride, Old Father Coyote who makes and unmakes the world in a long cycle of stories and, incidentally, gets his penis caught in a cleft pine for his pains. (*Hours* 11)

Coyote, the trickster, tricked people, but they also tricked him. Life was a gamble for primitive humans and they played with cunning and skill, but with the sense their adversary was cunning and skillful, too. In traditional Native American mythology, the trickster usually assumed the guise of Coyote, Raven, or Hare. Eiseley's choice of the rat as trickster seems to come from personal experience, not from an attempt to adopt some Native American mythological system whole. The childhood friend called "the Rat," the boy who taught Eiseley the dangerous subterranean ways of the sewer and died suddenly, may have been the initial trickster figure in Eiseley's life, as well as in his work, but Eiseley also used the rat elsewhere as a trickster, particularly in "Big Eyes and Small Eyes," (*Gentry*) where a rat watches an outdoor dinner party from under a chair and Eiseley notices the ironic contrast between the white-clothed man in the chair pontificating philosophically and the rat beneath the chair.

The pack rats play an important part in Eiseley's autobiography; they were his only companions for much of the time he was recuperating in the Mojave desert; they stole his glasses and his watch – certainly good metaphors for the way in which the trickster element in the universe steals time and foils the human attempt to see clearly. As he states in "The Time Traders" chapter of the autobiography, "the brain itself is a rat midden." The trickster is within ourselves, as well as outside.

Eiseley's explanation of the trickster element in the universe, in ourselves, is complex. In his initial explanation of the trickster in *All the Strange Hours*, he says:

> While I tossed sleepless, waiting for morning, the rat danced at the periphery of my vision, an afterimage that could not be exorcised – not till long, long afterward. The trickster who humbles pride. Every funeral has one. Only the old people from the silent cliff houses and the horse people of the plains had known and institutionalized him – the backward dancing man, the caricaturist of order. As for me, the rat went on dancing, dancing in my wearied brain as I rolled upon the pillow. *Behind nothing, before nothing, worship it the zero.* (*Hours* 12)

The trickster has to do with death and time, and Eiseley identifies the rat with the trickster, but he evidently does not want to give his beloved animals the "heavy" role in the drama, for, as the autobiography continues, the Other Player takes on many of the more sinister aspects of the trickster figure. The Other Player is an apt metaphor, reflecting not merely Eiseley's own dice thrown in childhood's abandoned house, but the cultural habits of primitive men. Even tribes like the Naskapi, who never even adopted the sophisticated, though simple, protective clothing and tools of their Eskimo neighbors, had still developed a great variety of games, some of them games of chance and gambling. The Naskapi, like other hunting tribes, enjoyed their games but also took them seriously as magical dealings, pastimes with religious implications. No doubt they understood how such pastimes reflected the gamble of the hunt, the gamble for survival against the cold and hunger of Newfoundland and Labrador (Speck 200–204).

In the closing paragraphs of the book, "Coyote the trickster who is unscrupulous and wins at gambling," as well as the trade rats who keep the records of time in the collections of things in their middens, would help man and dog as they moved further and further back in time. Coyote is the gentler gambler from an older culture who will help Eiseley and Wolf past the Other Player, the character with whom Eiseley gambles over life, the dominant trickster figure in *All the Strange Hours*. *All the Strange Hours* is the story of the gamble called life, an attempt to understand what the game is and why one man played as he did. It is also an attempt to learn to live with the adversary, the Other Player.

Eiseley probably emphasizes Coyote as a gambler at least partly because of an early memory. On this occasion, the young Eiseley was exploring an abandoned farmhouse, where he found a paper with the name "Eiseley" on it and a pair of dice. As the sun set, the child played, not a game he learned from others, but his own game. And he says of that time:

> I think I played against the universe as the universe was represented by the wind, stirring papers on the plaster-strewn floor. I played against time, remembering my stolen crosses. I played for adventure and escape. Then clutching the dice, but not the paper with my name, I fled frantically down the leaf-sodden, unused road, never to return. (*Hours* 29)

Eiseley had come early to a sense of death. As a small child, he had buried dead birds and clippings about war aces who were killed, tenderly making little gilt crosses to mark the graves. Someone mowing the empty lot had stolen the little crosses, and he mourned the dual death that occurred when first the body, then the memory or memorial disappeared.

The shadowy adversary that was time and death and evil felt very real, not merely in the abandoned house: Eiseley felt it walked the aisles of the hatchery in which he worked, somehow connected with the dead and deformed chicks and the sudden fire that once almost trapped and killed him; it was the figure who confronted him in a dream and became, when Eiseley got his hands on the invisible figure, a shriveled, grinning, papier-mache puppet; it was the cowled figure at the campfire whose hood covered a black emptiness instead of a face; it was the thing with rifles stalking him through endless snows to gun him down; it was the rat who danced on the platform as he gave a speech in "the first year I ever thought of myself as old" (*Hours* 4).

Eiseley saw life as a gamble, himself as a man "whose profession, even his life, is no more than a gambler's throw by the firelight of a western wagon" (*Hours* 24). And in life itself, Eiseley saw not merely what was, but the potential of all that was not: "I had learned that the machinery of life is gambling machinery, bringing into existence both the beautiful and maimed. I had seen the reality in all its shapes, like dice throws on a green table" (*Hours* 120).

We always lose. That is the nature of the games we play with the Other Player. We cannot win, but we can gain the cosmic distance that lets us see our little game of life and death as a number, and time as the succession of zeros that stretches behind it, comes after it. "Worship it, the zero" the student of time counseled as he moved backward through the snow.

How to face death, the death of individual beings, the death of a species, and the possible disappearance of life itself are grave concerns in Eiseley's autobiography. He connects evil strongly with death; in the earlier works he identified man's destruction of other life forms with evil, and saw his own humanity, goodness perhaps, in caring for "the lost ones, the failures of the world" (*Universe* 86). In the work where that statement appears, "The Star Thrower," Eiseley explains his ideas of dualism clearly:

> A hidden dualism that has haunted man since antiquity runs across his religious conceptions as the conflict between good and evil. It persists in the modern world of science under other guises. It becomes chaos versus form or anti-chaos. Form, since the rise of the evolutionary philosophy, has itself taken on an illusory quality. Our apparent shapes no longer have the stability of a single divine fiat. Instead, they waver and dissolve into the unexpected. We gaze backward into a contracting cone of life until words leave us and all we know is dissolved into the simple circuits of a reptilian brain. Finally, sentience subsides into an animalcule. (*Universe* 75–76).

Thus, Eiseley sees the issue of good and evil as distinctly subordinate to that of life and death, and both as matters subordinate to time – the time in which life evolved out of nothing, the time that would take us back to nothing, the zero behind and in front of us – the god scientific humanity, like the ancient Maya, could worship.

E. Fred Carlisle

The Literary Achievement of Loren Eiseley

In October, 1947, Loren Eiseley published an essay titled "Obituary of A Bone Hunter" in *Harper's* magazine. This presumably auto-biographical sketch related three incidents from the 1930s in which Eiseley failed to make the big find that would have made him famous. The first occurred in a cave near Carlsbad, New Mexico, when spiders turned him back. The second missed opportunity happened in a similar cave, and that time he turned back because of an egg in an owl's nest. The third involved an apparently crazy old man and a fragment of what might have been a fossil human jaw bone. Eiseley attributed these failures to the "folly of doubt," rather than to bad luck or limited scientific ability, and he concluded his account of "the life of a small bone hunter" by saying, "I have made no great discoveries." "There will be no further chances."

The finality of that seems odd coming from a man who had turned forty only a month before the essay appeared and who was on his way back to Penn to assume the chairmanship of the Anthropology department. But evidently, it continued to express something of Eiseley's belief about his career, for twenty-four years later, he included the "Obituary," virtually unchanged, in his autobiographical collection, *The Night Country*. "I have made no great discoveries." Well, in a strictly scientific sense, perhaps he did not. But as an estimate of his career, there is a great irony in that sentence, for Loren Eiseley did, in fact, make an extraordinary discovery when he came upon his true and original genius in the unique voice and vision of *The Immense Journey*.

In what he considered his favorite book, Eiseley told the story of human evolution, but his story was not science in the usual sense. He also included "a bit of his personal universe," and so the book became, as well, "the record of what one man thought as he pursued research and pressed his hands against the confining walls of scientific method in this time." In this unconventional record, Eiseley sometimes spoke extravagantly about miracles, the magic of water, and enormous extensions of vision. Like Thoreau, Eiseley wandered beyond certain professional and literary orthodoxies and tried to speak the truths he had discovered. As he said later in *The Unexpected Universe*, "I am trying to write honestly from my own experience." Like Emerson, Eiseley searched for an adequate geometry – a kind of new idiom or language – to express the mysterious and contradictory nature of experience that science had helped him discover.

With the 1957 publication of *The Immense Journey*, Eiseley completed a critical phase of his literary development. He had found a way to intertwine all the strands of his personal and professional life into a coherent whole. *The Immense Journey* tells one of the epic stories of science and simultaneously provides an imaginative exploration and an artful expression of it. In short, Eiseley intertwined autobiographical, scientific, figurative, and metaphysical elements into a new idiom and a unique vision, and that is the heart of his literary achievement.

The voice and vision were both firmly grounded in science. Without that experience and knowledge – and this is as necessary as it is obvious to say – Eiseley could never have become the original and accomplished writer he was. The facts of his scientific development and career are well known and require no summary here. His scientific *achievement* is less certain, but it was clearly substantial. Perhaps he did not succeed as either a researcher unearthing important new fossils and artifacts or as a theoretical anthropologist advancing and supporting new formulations for inadequately explained data. Nevertheless, he participated in field research of some significance; he had a remarkable command of his and related fields of knowledge; he published a great many important analytical and critical papers; he developed a great ability for exposing the inadequacy of other's work; he was in fact an excellent scholar and critic in his field – and later a significant historian of science. Eiseley became a very successful academic scientist, and that in itself is a notable achievement.

But even more than that, science shaped Eiseley's thinking about nature and human experience, and his writing about science influenced his later literary style. It is fascinating in itself, but not the point here, to trace his development as a writer of science and to see the modulation from a very conventional professional style toward the new idiom. But the most obvious literary foundation for his achievement, however, was the poetry and prose he published in the thirties and early forties.

In the ten years since his death, this earlier work has become better known and like the facts of his scientific career needs no lengthy review here. What is important to remember is that, at least indirectly, the work is autobiographical; it is firmly tied to place and to the historic and prehistoric pasts; and it reflects a questioning if not yet scientific mind. The poetry bears the mark of a distressing childhood and youth, as well as young adulthood, and it reflects Eiseley's attachment to place – the spaces extending from the prairie around Lincoln to the uplands of western Nebraska and beyond. Besides its personal and confessional quality, the poetry speaks, as well, of an entire region and people for whom struggle, poverty, and failure were a way of life.

As a young poet Eiseley developed considerable control, complexity, and subtlety in his work and achieved a level of sensitivity and maturity that extended the significance of his poetry beyond the experience of a particular young man and region. He wrote well in the mid-1930s. Nevertheless, his work created no new poetic idiom, nor did it discover any new perspectives. The images, forms, and themes were all rather conventional. The development in his style reflected to some extent his increasing personal confidence and stability, but Eiseley did not discover in poetry a way of moving beyond certain emotional and psychic resolutions or beyond certain literary conventions.

In science, by contrast, he seemed to find the order, system, and ideology he needed to insulate himself from the turmoil of his life. There, he discovered control and explanations that confirmed his experience, gave it wider significance, and enabled him to escape the relative isolation of a lyric poet. As a member of a professional community, he developed a style appropriate to the impersonal and analytic conventions of science. But that style then gave way to a more personal and fluent language. Eiseley was apparently trying to recover certain qualities of expression, feeling, and insight that he had developed as a poet but subdued as a young scientist, and was trying also to expand the potential meanings of his prose. This development – and it is apparent by the late forties – led in just a few years to what I am calling a new idiom.

This new mode or style can most clearly be characterized through the layers or dimensions Eiseley so artistically intertwined in *The*

Immense Journey: science, autobiography, figuration, and metaphysics. The layer of science is very clear. It is realized in the story of evolution that Eiseley tells – a direct, informed account that gives the book its narrative and expository structure.

The autobiographical aspects are equally clear. From the first chapter on, Eiseley is telling two stories – one about evolution, the species history, and one about himself from his personal experiences as a fieldworker to his discoveries of significance and relationship. Autobiography constitutes part of what he calls the "unconventional record" of the book.

Figuration or metaphor is, first, the dimension that gives interest, texture, and impact to Eiseley's style. Such dramatic metaphors as "out of the choked Devonian waters emerged sight and sound and the music that rolls invisible through the composer's brain" compress time and radically connect different species in a powerful and meaningful way. Through them, Eiseley helps us see *our* place in the story – not just as its end but as beings still connected to its origins. The metaphoric dimension also enables him to relate the two histories. The "slit" which Eiseley describes in chapter 1 is but an example. Actual place, occasion for explanation and speculation, and metaphor for time and for Eiseley's situation in time, the "slit" helps intertwine all of the dimensions.

The metaphysical or speculative layer is associated most closely with the enormous extension of vision Eiseley speaks about. Although the later chapters seem more speculative than earlier ones, this layer is present throughout, as Eiseley's meditations about himself and time in the first chapter suggest.

In *The Immense Journey*, and in *The Firmament of Time* and *The Unexpected Universe*, Eiseley has made explicit and functional aspects of discourse and of experience that are at best implicit or invisible, and often absent, in the writing of most scientists or poets. This alone suggests that he is not simply a scientist who writes well about science or about human values; nor is he simply a popularizer (if he is one at all) who explains science to others; nor is he just a writer who makes "poetry" out of science, thereby changing it into something else. His achievement of a new idiom is more considerable.

In 1974, I characterized Eiseley's vision as a "heretical science," and while his way of thinking about science may seem less unusual today, there is no reason to think differently about the unorthodox character of that vision. I am not speaking about the themes of Eiseley's books and essays or of his ideas. Vision is something more basic; it informs and shapes ideas and themes. In the simplest sense, Eiseley discovered in *The Immense Journey,* and elsewhere, a greater and more comprehensive version of science than the usual or orthodox sense. It was in the words of Francis Bacon, a science "for the uses of life," and in a personal sense, it became a science for the uses of the self. It is science which *includes* the self.

Eiseley accepted evolution as a well substantiated scientific theory. His own professional work helped confirm or elaborate the theory and he studied and learned far more than he was able to observe directly. However, he also interiorized the theory, so that it functioned as a major structure for perceiving and comprehending experience. He dwelt in it, so to speak, and through it he made contact with reality. His research and travel, his scientific knowledge, and his belief in the modern theory of evolution gave his work perspective, shape, and authority, as well as content.

While Eiseley believed that scientific knowledge helps us understand the world, he could not forget that we are also often baffled by the natural world and limited and misled by our incomplete knowledge of it. Science liberates us with its knowledge and vision, yet it repeatedly redefines our limitations as it raises new problems and reveals new boundaries. He accepted the systematic structure of scientific method and knowledge, and they took him a long way. At the same time, he pressed against the narrow, confining limits of the very system he accepted, as he struggled to see through to the meanings hidden in nature. He tried to extend science so that it comprehended more, but he also tried to understand for himself. So no matter how far the instruments and structure of science carried him into nature, Loren Eiseley was still, in a sense, tracking himself – still seeking to establish *his* reality.

As a professional anthropologist, Eiseley searched widely for the origins of humankind. But that search extended naturally and logically to a search for the origins of himself in the fields and small towns of Nebraska and in the uplands of the ice age. It widened to a quest for himself as a solitary fugitive from the twentieth century and as one of a lonely and wandering species through time. It concentrated into a search for his inner self in the dark void and deep spaces within himself and in the evolution of the human brain – "in the windswept uplands of the human mind." Eiseley reached into the depths of time, of the earth, of life, and of himself – he classified *and* contemplated – in search of a more comprehensive science.

In *The Firmament of Time,* he writes " 'The special value of science,' a perceptive philosopher once wrote, 'lies not in what it makes of the world, but what it makes of the knower.' " Loren Eiseley, however, did not reject one for the other. The one – the world – leads into the other – the self – for in one sense the self includes the world or holds some of the answers to it: "Man's quest for certainty is, in the last analysis, a quest for meaning. But the meaning lies buried within himself rather than in the void he has vainly searched for portents since antiquity." The value of the scientist's activity and knowledge rests in *their* value and meaning for the lived life – "for the uses of life."

Several writers believe that Eiseley's achievement is limited primarily to the personal or familiar essay – or to what he called the concealed essay. They quite reasonably associate Eiseley with the natural history essay in America and with authors like Henry Thoreau, John Burroughs, and John Muir, and, in the twentieth century, Aldo Leopold, Rachel Carson, Annie Dillard, and Lewis Thomas. There is no question that Eiseley was influenced by early nature writers, as he was by G. K. Chesterton. And Eiseley has made a significant contribution to the art. He must indeed be read and valued as an essayist. Nevertheless, his achievement exceeds that genre. Like Thoreau, Eiseley wrote at least one extraordinary book in which the artistry and meaning surpass the limits of the essay. In *The Immense Journey* certainly and perhaps in *The Unexpected Universe* – at his most successful in other words – Eiseley fused his essays into unified, coherent books in which the whole is greater than any sum of its parts.

And while Eiseley's finest books and essays certainly achieved what he attributes to the great nature writers, his vision nevertheless also exceeds theirs. In writing about Sir Francis Bacon, Eiseley said that "Words can sometimes be more penetrating probes into the nature of the universe than any instrument wielded in the laboratory." In saying this about Bacon, Eiseley seems to mean that words can be *scientific* instruments.

Words give the scientist a means to explain and interpret research and discoveries, and even here they are not altogether transparent. They are a medium which shapes the scientist's insights and therefore affects his meaning. But Eiseley means even more, for words function sometimes as the *primary* instrument of the scientist; they become the medium within which the scientist's knowledge and imagination operate in order to discover something of the nature of the universe. Words, I believe, functioned this way, scientifically, for Loren Eiseley.

Words also provide the great nature writers with a powerful instrument of insight and expression. Their contributions, however, are different. They add a dimension to science, "something that lies beyond the careful analyses of professional biology." Science requires their sensitivity and insight, for without it "we are half blind . . . we . . . lack pity and tolerance." These nature writers provide, apparently, a kind of synthetic vision. They write about nature *and* humanity. They see nature through frankly human eyes and try to express its beauty and meaning in human terms.

Now, certainly Eiseley did that, and in writing about the great nature writers, he may have been making a case for himself. But it is mistaken, I believe, to regard him simply as a nature writer who understood and valued science – when in fact he was a scientist, and he was also a writer – an artist. Because of this dual identity, his literary achievement was distinct.

In *The Man Who Saw Through Time* Eiseley distinguishes between the scientist and the artist in their great moments of creativity. Even though both may experience the aesthetic joy of discovery and design, "a substantial difference still remains. For science seeks essentially to naturalize man in the structure of predictable law and conformity, whereas the artist is interested in man the individual." Eiseley believes that a person may be both a scientist and artist, but somehow the two still remain separate; their activities do not fuse into a single mode of creative activity. While the discoveries of the great artist or scientist might give us a "new geography" – a new design for reality which in turn changes our world, their domains for Eiseley are nevertheless distinct. The artist explores the interior – and draws the world and us within; whereas the scientist sometimes ventures to "remake reality."

The distinction seems sound enough. But perhaps in Eiseley's own case the two ventures are not so different as he thought. Perhaps, *both* the scientist and the artist discover and remake reality. The scientist may concentrate on the physical and chemical or organic and physiological constituents, and the artist may dwell on the emotional and psychological constituents of reality. In our time, however, these domains are not always so easy to separate, and sometimes they veer too close together for intellectual comfort.

Eiseley tried, I believe, to combine science and art as he explains them. He concentrates on the individual and on humanity within the structure of predictability (and unpredictability) defined largely by anthropology and biology. In his writing the systematic activity and structure of science merge with the search for the self within that structure. Science, then, is simultaneously a pursuit of the self and an attempt to make increasingly closer and closer contact with reality. The personal and universal dimensions of science merge in Eiseley's books, as the quest for knowledge of reality and knowledge of oneself become one.

Eiseley realizes his unique voice and vision most fully in *The Immense Journey* and in *The Unexpected Universe*. They are the heart of his literary achievement. Nevertheless, he wrote two other impressive books – *Darwin's Century*, his major work in the history of science, and *All the Strange Hours*, his remarkable autobiography. Both are memorable, and from my point of view, they stand as complements to the books I value most. The preparation of *Darwin's Century* certainly influenced and informed *The Immense Journey* and represents the scholarly and scientific aspect of Eiseley's work. The autobiographical and figurative dimensions are clearly subdued. In *All the Strange Hours*, Eiseley subdued the scientific dimension of his work in order to excavate his personal life. Without these two books, Eiseley's literary achievement would be incomplete, certainly, but each in itself reveals only an aspect of his original genius. Without *The Immense Journey* and *The Unexpected Universe*, there would be no original – no literary achievement for the ages.

Peter Heidtmann

An Artist of Autumn

Where are the songs of Spring? Aye, where are they?
Think not of them, thou hast thy music too . . .
– Keats

In *The Fall* Camus's mordant narrator refers to a "man who, having entered holy orders, gave up the frock because his cell, instead of overlooking a vast landscape as he expected, looked out on a wall." We quickly grasp the unstated idea. How could the monk achieve the goal of enlarging his soul to accommodate the divine if his vision was so closely confined? The theotropic mind often requires an analogue in the external world to help it explore the metaphorically vast inner regions where it hopes to find the immanent god.

Although it is unlikely that any established orders would have been suited to Loren Eiseley's contemplative inclinations, there can be no doubt that he had a mind of this sort. And since the forms of religion arise in response to certain spiritual needs, it should not be surprising that the needs remain despite a loss of vitality in the forms that once satisfied them. Thus Eiseley, who comes across in his books as a man of almost monkish isolation, looks out at the world with a yearning that is perhaps even more intense than that experienced by solitary seekers in less anxious centuries. For he is a representative modern man, that figure (as he puts it) "of ruinous countenance from whom the gods have hidden themselves."

Yet this situation does not deter the pilgrim soul in Eiseley. Instead he asserts that human beings, having been deprived of instinct, *must* search for meanings. Nowadays it is customary to distinguish three broad fields of endeavor – religion, science, and art – within which this search is conducted. But all three have their origin in "a common instinct of wonder," as Richard Carrington has pointed out, and were constituents of "a single mental process" in primitive thought. In considering Eiseley's own quest as contained in his essays, we become aware of a survival (or recurrence) of this outlook.

For one thing, despite a pervasive melancholy, Eiseley never lost his sense of wonder. Again and again in his writing he prompts us to remember – to *see* – that nature "is one vast miracle transcending the reality of night and nothingness." It is as if he himself had come from elsewhere and were continually amazed by his surroundings on this planet. As he puts it, "For many of us the Biblical bush still burns, and there is a deep mystery in the heart of a simple seed."

Secondly, if we reflect upon his essays as a whole, we realize that science, religion, and art are not compartmentalized in them, that the three seem instead to be aspects of a single mental process. By profession he was a scientist, an anthropologist who devoted special attention to the theory of evolution. At the same time he was aware that modern science, for all the sophistication of its methods and instruments, could not get below a certain depth in its efforts to explain the physical world. The enigma of life would always elude its grasp. Even in his first book, therefore, he mentions having found the scientific method somewhat confining, and in a later essay he acknowledges having turned away from some of the logical disciplines to which he adhered earlier in his career.

In place of them Eiseley adopts the less rigorous approach of the literary naturalist. Contrary to the laboratory scientist, for whom the phenomena of the universe are objective facts that he must try to understand by discovering their underlying patterns or principles, the naturalist makes no effort to distance himself from the objects of his study. His more personal outlook suggests that "there is a natural history of *souls,* nay, even of man himself, which can be learned only from the *symbolism* inherent in the world about him" (my emphasis). The two italicized words are not part of the scientific vocabulary. By using them here Eiseley shows his willingness to enter the domain of religion, and even to obscure our customary distinction between religion and science.

What of his art? Since without it the essays would not exist, we must ultimately see it as the primary instrument of Eiseley's lifelong quest. It is not only the vehicle for the conveyance of his religious and scientific outlook, but it is also the means by which (in varying degrees) he lures us into participating in his vision. Eiseley, it must be admitted, is not especially noteworthy for any original insights into the problems examined by scientists and theologians. Instead it is his evocative way of treating them that is most memorable. In effect, then, Eiseley the autobiographical essayist functions chiefly as a poet in his prose.

His own awareness of himself as an imaginative writer is revealed in part through explicit references to his storytelling role. In introducing an anecdote from his school days, for instance, he writes, "Yet bear with me a moment. I would like to tell a tale, a genuine tale of childhood." And as a lead-in to an episode set during one of the seasons he spent in the field as a fossil hunter, he says, "Suppose that there still lived . . . but let me tell the tale, make of it what you will." In fact Eiseley is an irrepressible teller of tales. Even when it is not his main intention to narrate part of his life story, he frequently uses little stories from his life for purposes of illustration. His essays are rich in personal anecdotes.

Yet he is more than merely a rather old-fashioned raconteur; he also reveals a penchant for myth-making. This is suggested in a passage from *The Unexpected Universe* where, delving deep into his own sensibility, he writes:

> Man, for all his daylight activities, is, at best, an evening creature. Our very addiction to the day and our compulsion, manifest through the ages, to invent and use illuminating devices, to contest with midnight, to cast off sleep as we would death, suggest that we know more of the shadows than we are willing to recognize. We have come from the dark wood of the past, and our bodies carry the scars and unhealed wounds of that transition. Our minds are haunted by night terrors that arise from the subterranean domain of racial and private memories.

Eiseley's memories of childhood and youth, collected principally in *The Night Country* and *All the Strange Hours*, provide the private or individual basis for his development as an isolate haunted by darkness. Consistently presenting himself as a wanderer through "the wilderness of a single life," he comes across in his work as a modern Ishmael. The transpersonal sources of this mythic image are less easily identified, but they can be understood as welling up from the pool of common human memories.

Eiseley's access to that subterranean domain is connected with his ability to penetrate (in Shakespeare's phrase) "the dark backward and abysm of time." This is partially a result of his learning what scientists have discovered about our evolutionary past. But his scientific knowledge seems to have aroused in him a deeper, pre-rational apprehension, which sometimes flashes out in sudden, startling perceptions. In one place, for example, he mentions that he has sat listening to a great singer in a modern concert hall, and at the same time has "heard far off as if ascending out of some black stairwell the guttural whisperings and bestial coughings out of which that voice arose." Elsewhere he describes his disturbing vision, while standing on a lecturer's rostrum, of a great tree trunk stretching loathsomely behind him along the floor. Trembling there with book and spectacles, he realizes that he is himself "a many-visaged thing that has climbed upward out of the dark of endless leaf-falls, and has slunk, furred, through the glitter of blue glacial nights."

Indeed Eiseley could not forget that man had emerged during the Pleistocene, that (in a manner of speaking) "the ice had made him." His own breath partakes of the breath from that dank door, the coldness released by it mixes with the marrow in his bones, and the darkness from behind it haunts his mind. Especially the darkness, which he also refers to as the night tide, "because that is the way you come to feel it – invisible, imperceptible almost, unless it is looked for – and yet, as you grow older you realize that it is always there, swirling like vapor just beyond the edge of the lamp at evening and similarly out to the ends of the universe."

Although Eiseley was drawn to the dark, he admits that, like the rest of humankind, he feared it as well. For him that fear is most particularly embodied in rats, but the dread of darkness need not assume external form. It can also take shape solely in the mind because of our knowledge that for each of us one night will be unending. This knowledge, which distinguishes human beings from all other animal species, is in itself fearful. Primitive man tried to cope with it by means of magic, and Eiseley – like all artists – can also be seen as a magician. He has a religious sensibility and a great deal of scientific understanding, but in the essays verbal magic is his ultimate resource. In this sense he is like the primitive caster of spells, hoping through words to keep at bay the powerful forces of death and disorder that threaten to erupt from the surrounding dark.

There is no doubt in Eiseley's mind that those forces will finally triumph. Yet in confronting the endless night they represent, he does something unusual. Because life for him is "a magnificent and irrecoverable good" which exists for only a short while within each of us, he celebrates autumn, life's last stronghold against the on-slaught of wintry darkness and death. Autumn is his season, and he acquaints us with its music.

It causes no surprise to learn that the collection of poems Eiseley was working on at the time of his death is entitled *Another Kind of Autumn*, but it does seem remarkable that the autumnal theme should be announced in his very first piece of published prose. It is a sketch of less than two pages called "Autumn – A Memory," and appeared in the initial volume of *Prairie Schooner*. At the time (1927) he had just turned twenty, but he was already acutely aware of time's passing. In the sketch he meditates upon a solitary visit he presents himself as having made to an Aztec ruin during some previous fall. Taking us back with him to the stillness of that fading afternoon, he remembers images of the long dead Indians who had once lived among the crumbling walls: "There were visions of the laboring copper bodies that built this place under the blazing sun. . . . There were generations, there were brave little friendships, hatred and feasts and there was love." He also wonders, with no hope of being answered, how those people all came to die – whether in battle, or because their harvests burned, or through the displeasure of some offended god. A few lines from the end, in a sentence that anticipates his mature style, all he can do is acknowledge that "I was a shadow among shadows, brooding over the fate of other shadows that I alone strove to summon up out of the all-pervading dusk."

Eiseley's preoccupation with this season is deeply rooted in his own life, for he never forgot that he was a belated son, born when his father was forty, and the offspring of an unhappy marriage. These conditions of his origin lead him to say that he was "an autumn child surrounded by falling leaves." The implication is that destiny was at work, that he was a man marked for endings from the very start. In any event he dwells on the autumn theme at or near the end of three of his books of essays – *The Immense Journey*, *The Unexpected Universe*, and *All the Strange Hours*.

In the concluding chapter of the first of these, published when he was fifty, Eiseley imagines himself taking an autumn walk. Wearing a hat and an old jacket, he goes outdoors on a day "when the leaves are red, or fallen, and just after the birds are gone." After climbing over a wall, he walks across "an unkempt field full of brown stalks and emptied seed pods," then comes to a wood where he finds a place to rest and consider "the best way to search for the secret of life." Eiseley's writing here calls self-conscious attention to his stance as an old-fashioned naturalist motivated by a loving curiosity to explore the visible world in hopes of getting closer to its invisibly beating heart. He considers the season appropriate to his purpose because hectic activity and green leaves are not present to confuse the issue. Instead, as he writes, "The underlying apparatus, the hooks, needles, stalks, wires, suction cups, thin pipes, and irridescent bladders are all exposed in a gigantic dissection." Because these essentials provide "an unparalleled opportunity to examine in sharp and beautiful angularity the shape of life," he determines from henceforth to give autumn his "final and undivided attention."

Unlike the early sketch, which is permeated by a youthful *Weltschmerz*, this passage reflects the attitude of one who is aware of growing older and wishes to conserve his outwardly directed energies. In the other two autumnal passages, however, Eiseley's attention takes a more inward turn. That in itself would result in a tonal change, but in addition both of them were published when he was much closer to the end of his life.

The earlier of the two, part of the last chapter of *The Unexpected Universe*, came out when Eiseley was sixty-two. He begins the essay by pondering the "slow-burning oxidation" in the brain that enables it to hoard memories, and he is subsequently reminded of a wild-plum thicket in Nebraska that he visited one autumn during his youth. The fallen plums contained a simpler version of "the mystery hidden in our heads," he writes. "They were hoarding and dispersing energy while the inanimate universe was running down around us." On the last pages of the chapter he comments again on the plum thicket, this time in connection with a second visit he made to it on an autumn walk years later. He returned partly just to see if it was still there, but also because he was still puzzled by the "strange hoarding and burning at the heart of life."

This much is reminiscent of the passage from *The Immense Journey* in that it suggests a continuing concern with the mysteries of the organic world. But the mood changes when he makes a different connection between the plums and himself. With his head grown heavy and the smoke from the autumn fields seeming to penetrate his mind, he says he felt like dropping all his memories just as the fruit fell about him from the trees. His desire was "to strew them like the blue plums in some gesture of love toward the universe all outward on a mat of leaves. Rich, rich and not to be hoarded, only to be laid down for someone, anyone, no longer to be carried and remembered in pain." In this frame of mind he leaned farther back into the leaves and was overcome by a strange yet soothing feeling, one he had never had before. "Perhaps I was no longer *Homo sapiens*," he imagines. "Perhaps all I was, really, was a pile of autumn leaves seeing smoke wraiths through the haze of my own burning." Letting go, he had the sensation, in other words, of being no longer partially separate from nature but of returning wholly to it. Although a harbinger of the terminal loss of consciousness, the experience in this context is far from frightening. Instead it is reminiscent of another familiar line from Keats, in which the speaker admits that many a time he has been "half in love with easeful Death."

The final passage is contained in the last of Eiseley's books to appear during his lifetime. Published when he was sixty-eight, *All the Strange Hours* is subtitled "The Excavation of a Life." In it he turns to an explicit consideration of the autumn theme a few chapters from the end, being concerned this time to demonstrate its applicability to his life within the context of his family background.

He begins by referring to the dreams that plague him, some of which prompt him to strike out fiercely in the dark, while others cause him to weep or pant in fear without apparent cause. He also mentions his bouts of sleeplessness, during which times the persistent coming and going of pictures in his brain is beyond his control. Then he adds,

> This is the beginning of age as all my family have known age: my grandmother Corey, who periodically cried out desperately in her sleep for help but who, upon being awakened, never confided what it was she feared; or my mother, who finally stalked the dark house sleepless at midnight; or my father, who in the great influenza epidemic of 1917–1918 came home unassisted, went to his room, and lay quietly for days without medical attention, only his intelligent eyes roving the ceiling, waiting for which way the dice would fall and not, I believe now, caring overmuch.

He explains all these behaviors as final efforts on the part of the individuals mentioned to order the meaning of their lives. This, he observes, is "the human autumn before the snow."

If spring is the season of bright promise, then fall is the time of disillusionment. Hope, the expectation of some fancied fulfillment, gives way to anguished recollection. As Eiseley puts it, "Oncoming age is to me a vast wild autumn country strewn with broken seedpods, hurrying cloud wrack, abandoned farm machinery, and circling crows. A place where things were begun on too grand a scale to complete." The imagery would not be the same for everyone, of course, but he thinks that the essential feeling is universal. Consequently his intention in *All the Strange Hours* is "to bespeak, in some fashion, the autumn years of all men." And he suggests that they are years in which the individual, impelled to review the natural history of his soul, vacillates between apprehension and anticipation at the prospect of the pilgrimage coming to an end.

The apprehension is common enough, since no one can remain immune to dread when faced with extinction. Nevertheless Eiseley's treatment of it is enhanced through his planetary perspective. Because he knows that time extends millions of millennia behind us and lies incalculably far ahead of us, he offers the following advice in words that allude to one of his own poems: "Worship, then, like the Maya, the unknown zero, the procession of the time-bearing gods." Still he does not let his awareness of these vast reaches of time prevent him from seeing that each individual life, brief though it is, imitates the situation of the world at large. Billions of years ago it emerged from the void, and at some time in the unforeseeable future it too will return to sunless oblivion. Thus the darkness that exists both before and after our separate lives is a microcosm of the macrocosmic plight of the planet. Understood in this way, the autumn of Eiseley's life mirrors in miniature the prelude to the destruction of our universe.

The anticipation of the end is a different matter. Others have surely felt it, but Eiseley's way of handling it is unmistakably his own. The experience in the plum thicket provides one illustration of this anticipatory feeling; another is described in a still more striking account (contained in *The Star Thrower*) of his visit to a great hill in Montana.

After delivering a lecture in Concord, Massachusetts, Eiseley had flown west. There, while wandering over a sunbeaten upland, he picked up "a quartz knife that had the look of ten thousand years about it." He says that it was "as clean as the sun," and that suddenly he realized what Thoreau had been thinking about in regard to all the arrowheads he found scattered around the countryside. "They were free at last," Eiseley declares. "They had aged out of human history, out of corruption." Following this insight he was overcome with an exceptional feeling which he explains in this way: "I too had taken on a desert varnish. I might have been a man but, if so, a man from whom centuries had been flayed away. I was being transmuted, worn down." In this mood he lay on the ground and drifted off to sleep, all the while feeling that he was stiffening into immobility, "freezing into the agate limbs of petrified trees." Nor was there anything to fear. On the contrary, he says, "I sighed a little with the cleanliness of that release. I slept deep under the great sky."

Although we are told that this experience literally took place in

July, its atmosphere is as autumnal as that of the plum-thicket episode, and it helps us to understand how Eiseley's self-identification with the season preceding winter is a way he has of mythologizing his life: the mood of anticipation is related in an unexpected way to the archetypal pattern of innocence/fall/recovery-of-innocence. In Eiseley's case neither a resurgence of spring nor the belief in a resurrection is needed to complete the pattern. Instead the wished-for fulfillment is a consequence of his backward yearning, of his nostalgic longing for a return to some primordial situation where time and self-consciousness have no meaning. In other places he projects himself into a more primitive human condition in the ice-age world; here, in the episodes set in the plum thicket and on the hill in Montana, he imagines himself merging with non-human and even non-organic matter. Thus we can see that, no matter what the mood, Eiseley's handling of the autumn theme is a manifestation of his need as an autobiographer to place his individual life within a larger context of signification.

In one of his *Texts for Nothing* Samuel Beckett has his monologist say, "[N]o need of a story, a story is not compulsory, just a life, that's the mistake I made, one of the mistakes, to have wanted a story for myself, whereas life alone is enough." It would be possible to argue that this attitude represents an extreme form of modern heroism on the grounds that extraordinary courage is required to face life without a story. On the other hand, because of its absence of meaning, story-less living can also be seen as scarcely human. As Eiseley has written, "No longer, as with the animal, can the world be accepted as given. It has to be perceived and consciously thought about, abstracted, and considered." To give up on this endeavor, to no longer strive to make sense out of the welter of phenomena with which we are confronted, is therefore to relinquish an essential part of our humanity. Despite his awareness that "life was never given to be bearable," Eiseley refused to abandon his own search for meaning. Instead, as with other autobiographical writers from the days of St. Augustine to the present, he struggled to assert his humanity by telling stories – and ultimately a Story – about himself.

One of his little narratives in *All the Strange Hours* concerns his childhood whittling of small wooden crosses that he painted with liquid gilt. How he used them and what became of them is explained in these words:

"I placed them over an occasional dead bird I buried. Or, if

I read of a tragic, heroic death like those of the war aces, I would put the clipping . . . into a little box and bury it with a gold cross to mark the spot. One day a mower in the empty lot beyond our backyard found the little cemetery and carried away all of my carefully carved crosses. I cried but I never told anyone. How could I? I had sought in my own small way to preserve the memory of what always in the end perished: life and great deeds."

Eiseley's original account of this espisode is told to W. H. Auden, who responded by saying that "it was a child's effort against time." A similar effort was made by the adult Eiseley through the writing of essays, even though the deeds preserved thereby are not great in any traditional sense. The point is that he again sought to make a mark in defiance of the scythe-wielding mower.

Because of all the anonymous monuments that have been found throughout the world, Eiseley the archaeologist is well aware of the universality of mankind's desire to leave some sign of his temporary presence on the earth. He invites us especially to consider the huge heads – survivals of the lost Olmec culture – that can still be seen in the jungles of eastern Mexico. They typify for him the work of groups of artisans from all over the globe, each of which arose to place "its stamp, the order of its style, upon surrounding objects, only to lapse again into the night of time." Still we are probably more accustomed to think of this drive as it is manifested in the artworks of particular individuals. Michelangelo's "David," for example, is a mark in marble that in a way has enabled its creator to overcome the destructiveness of the reaper. The autobiographical artist is similarly motivated, but the material he has to work with is his own life and character. In a sense, then, the task that occupies him is the creation of himself.

As Avrom Fleishman has written, "[A]n autobiography does not represent or repeat a life but instead brings it into existence. 'Life' means nothing in the individual (as distinct from the biological) realm until it is told in life stories." Eiseley would surely have agreed with him. This is suggested on one of the pages of his own autobiography, where he remembers his father coiling his fist and making him shiver as he read:

He was a kinde of Nothinge
Until he forged himself a name.

Clyde Eiseley may actually have spoken these words from Shakespeare, but if so he would have declaimed the entire passage, which reads:

> He was a kind of nothing, titleless,
> Till he had forged himself a name o' the fire
> Of burning Rome.

In their original context the words are critical of Coriolanus, who, after attaining the height of power, refused to recognize his former friends. On the other hand Eiseley's abridgment of the lines gives them a positive turn and emphasizes their application to himself as an artist.

To put the matter plainly, we may assume that the man produced by nature and nurture in Nebraska felt himself to be a kind of nothing. Yet, through his writing, he succeeded in forging himself a name. Telling stories in quest of his personal truth, becoming the product of his own artifice, he is the lonely figure of autumnal consciousness who not only dominates the pages of his essays, but also assumes a magnified stature in our minds after we have put down his books. This is the Loren Eiseley remembered by all those who did not know him personally. And if the destiny to which he often deferred is kind, he is the man who will likewise be remembered by generations still to come.

Ben Howard

Loren Eiseley and the State of Grace

Loren Eiseley once described himself as "relatively obscure." At the time, we were chatting in his spacious office at the University of Pennsylvania. On his desk lay the rough drafts of his autobiography, scrawled and disheveled. Beyond them were the shelves of books and relics, the bones and shells that kept him company in that solitary place.

Dr. Eiseley was then in his sixties, a broad-shouldered, gray-haired man with a rich, well-modulated voice. He was successful and looked the part, and his self-estimate seemed overly modest – or even falsely modest. Yet in retrospect I have caught the drift of his remark, and it seems neither inaccurate nor insincere. A shrewd professional writer, Eiseley was wondering aloud whether the life story of a moderately famous paleontologist would appeal to the book-buying public. A man haunted by extinction, he was also wondering what might be remembered of him after his death – what fragment of his reputation or remnant of his work might be saved from oblivion.

Twelve years later, that conversation stays in memory, as I consider what, indeed, has been spared from oblivion, and what, if anything, the present generation of students knows of Loren Eiseley. Since its publication in 1957, *The Immense Journey* has sold nearly a million copies and has been translated into several languages. His later books have done less well, but together they form a formidable corpus, and Leslie Gerber and Margaret McFadden, authors of *Loren Eiseley* (Ungar, 1984), are probably right in predicting that some of Eiseley's essays will become classics of the form, and that Eiseley will take his permanent place in a tradition that includes Montaigne, Gilbert White, Chesterton, and Thoreau. Yet doubts persist, as well they might. Not long ago a well-known historian came to the university where I teach and delivered a jeremiad on the fate of the planet earth. That was once Eiseley's theme, and much of the lecturer's grim message resembled Eiseley's of thirty years ago. Yet, in the course of the evening, though Rachel Carson, Paul Ehrlich, and other prominent environmentalists were invoked, Eiseley's name was never mentioned. Sometime later, in response to a student's question, I advised her to "read Loren Eiseley." "Is that a person?" she asked. Apparently the name sounded like the title of a novel.

Professionals in the fields of archeology, anthropology, and paleontology will ultimately judge Eiseley's contribution as a scientist and historian of science. His claim to posterity as an essayist is another matter, and I suspect that much will depend upon the fate of the genre itself – the familiar essay (or "concealed essay," as Eiseley liked to call it), in which factual materials and subjective responses are brought into accord, and the personal voice plays a role equal to that of scientific data. Eiseley did not originate the form, but he did much to raise its stature, and he remains one of its consummate practitioners. If interest in the familiar essay continues to increase, under the aegis of Annie Dillard, Edward Hoagland, Robert Finch, and other postwar writers, the best of Eiseley's essays – "The Flow of the River," "The Slit," "The Brown Wasps," "The Judgment of the Birds" – may garner more readers than their author had reason to expect.

What that new generation of readers will find is not only a superb prose stylist and a humane naturalist but also a writer who could do one thing more convincingly and more movingly than any modern writer I can think of. I am speaking of Eiseley's gift for evoking, in one or two pages, the spiritual condition of man in a state of grace. Eiseley did not subscribe to formal religion, and he entertained no belief in an afterlife. The spiritual state he reported, and so persuasively dramatized in his essays, occurred within the precincts of the present moment, where, however fleetingly, the self could escape its confines and the mind could exercise what Eiseley called "the lonely, magnificent power of humanity," "the most enormous extension of vision of which life is capable: the projection of itself into other lives." To such moments Eiseley brought a historian's knowledge and an evolutionist's awareness of passing millenia. But their setting was always the here-and-now – the unexpected, unpredictable occasion.

"All the blessings," wrote John Wesley, "which God hath bestowed upon man are of His mere grace, bounty, or favor; His free, undeserved favor; favor altogether undeserved: man having no claim to the least of His mercies" (*Standard Sermons*, I). Such is the saving grace in Eiseley's reported experiences, the unmerited bounty which is the fruit of faith rather than conscious will. To those experiences Eiseley brought a skeptical, secular intelligence, but he also brought the readiness of faith, whether the occasion was an eye-to-eye meeting with a fox cub, or the miraculous revival of a frozen catfish, or an encounter with a "star thrower," or an experience he described (in *All the Strange Hours*) as "The Most Perfect Day in the World."

Perhaps the most memorable of these moments occurs in "The Judgment of the Birds," where he witnesses, "by chance," a "judgment of life against death." The scene is a glade, and the focus of attention is a raven with "a red and squirming nestling in his beak." Out of the woodlands comes a "soft complaint" – the cry of the nestling's parents. That cry is taken up by other birds, until it, in turn, gives way to something more mysterious and momentous:

> The sighing died. It was then I saw the judgment. It was the judgment of life against death. I will never see it again so forcefully presented. I will never hear it again in notes so tragically prolonged. For in the midst of protest, they forgot the violence. There, in the clearing, the crystal note of a song sparrow lifted hesitantly in the hush. And finally, after painful fluttering, another took the song, and then another, the song passing from one bird to another, doubtfully at first, as though some evil thing were being slowly forgotten. Till suddenly they took heart and sang from many throats joyously together as birds are known to sing. They sang because life is sweet and sunlight beautiful. They sang under the brooding shadow of the raven. In simple truth they had forgotten the raven, for they were the singers of life, and not of death.
>
> (*The Immense Journey*, p. 175)

"No man," Eiseley tells us, "sets up such an experiment." To be its chronicler – its eloquent witness – was a grace granted to Loren Eiseley. It may also prove his hedge against oblivion, his lasting stone in the flux of time.

Caroline E. Werkley

Eiseley and Enchantment

It is easy to remember that the seven letter word "Eiseley" means the seven letter word "Enchant." It should be thus listed in all dictionaries, along with "Bewitch," "Charm," "Delight."

If Loren Eiseley had lived in the world of primitive man, to which he felt an extremely close relationship, he might have been the first shaman, the priest or magician who gave medical, magical and religious guidance. A shaman, according to the *Columbia Encyclopedia*, "is thought to be able to change his form, talk with spirits and travel to the other worlds."

Millennia later, anthropologist-author Loren Eiseley, who loved and wondered about the marvels and beauties of the Universe, would write, "Imagine, for a moment, that you have drunk from a magician's goblet. Reverse the irreversible stream of time." Eiseley drank constantly and eagerly from such a goblet and was aware that magic was everywhere, even in the world where man-made structures often dominated the wonders of nature. Magic also existed in the recognition that "the human hand . . . has been fin and scaly reptile foot and furry paw."

Like the early shaman, Loren Eiseley lived in a world of enchantment, a world he discovered when he was a boy and loved for the rest of his life. All nature represented distant worlds, distant ages he could always see, with the creatures who inhabited them, as he saw the universe of ten million years when he descended into a ravine that he described in "The Slit," published in his first book, *The Immense Journey*. Eiseley realized that he and all mankind are "potential fossils still carrying within our bodies the crudities of former existences, the marks of a world in which living creatures flow with little more consistency than clouds from age to age."

It was the ghost world of memory that convinced Eiseley he lived in a world of enchantment closely related to all other forms of life. To Eiseley, birds were not merely delicate feathered creatures of the present, flashes of scarlet, blue and gold, but shadows of scaled dinosaurs of ages long past. He saw in his mind his tiny mammalian ancestors scooting from these beasts, emerging from the shelter of bushes only at night when dinosaurs or other predators would not pursue them.

It is no wonder that he felt protective of any squirrels, turtles or other animals wandering in a street who might be hit by a car, as well as lost dogs and cats. When he saw a lost dog running wildly in Pennsylvania station in Philadelphia while he was waiting for his commuter train to take him home, he suffered terribly knowing the poor dog should be rescued, and that he could not do it. Later Eiseley was happier to be able to rescue the lost cat he found one evening near his own home and named Night Country, the title of one of his books. The thought of animals being used for medical experiments saddened him: he may have been seeing himself dissected in a science laboratory. A minister from Oklahoma who admired Eiseley's love of animals, as portrayed in his books, wrote to him that he had named his own little dog "Eiseley." A generous contributor to organizations that aided animals, Eiseley received the Joseph Wood Krutch Medal from the National Humane Society in 1976.

One of the great loves of Eiseley's 20th century life was books – ". . . the tiny figures by which the dead can be made to speak from those great cemeteries of thought known as libraries." ("The World Eaters," *The Invisible Pyramid*). Of a scholar's personal library, Eiseley wrote in an unpublished scrap discovered in his desk after his death:

> "These are the places where the mind, moving among the strange disparate litter of the Universe, whips out invisible tentacles, drawing in these elements and combining them into new and sometimes wonderful concoctions which, in their turn, will be passed on to the shelves of other men. There they may lie for a century, or two centuries, for five centuries, until some other radiant mind picks them up and rearranges them once more. Books are like the spores that a strange fungus exudes in the night. Many are lost but others lie waiting in dark corners for the fertile moment to come again." ("Report of Loren Eiseley Collection," Caroline E. Werkley)

As the man who himself created "tiny figures" walked the thought-cemetery aisles of libraries and book shops, the ghosts of Thoreau, Bacon, Montaigne, Darwin, Sir Thomas Browne, Gilbert White, and George Santayana and others called out to him. In his own home and in his Anthropology department office at the University of Pennsylvania Museum, both filled with books he loved, Eiseley also visited with his ghost friends. In his office he had a most interesting ghost, Othniel C. Marsh, a famous paleontologist of the previous century, who surely hovered over Eiseley's many reprints of his articles and the two massive, signed copies of his book *Dinocerata: A Monograph Of An Extinct Order of Gigantic Mammals* which Eiseley had been fortunate to discover on his book-collecting journeys. These expeditions were his favorite pastimes in Philadelphia and in any city he visited, as were journeys to the University of Pennsylvania library where he often checked out a dozen books at a time. Late at night, when no humans were in either office or Museum, Marsh undoubtedly joined other ghosts whose signatures were in books and reprints in Eiseley's library, including, among others, T. F. Powys, George Moore, Leigh Hunt, Walter De La Mare, and Joseph Leidy, the founder of paleontology in the United States and a University of Pennsylvania professor in the previous century, for whom the Joseph Leidy Laboratory of Biology at the University of Pennsylvania was named.

Other interesting ghosts in Eiseley's Anthropology department office included *Australopithecus africanus*, represented by a plaster statue that once stood in the University Museum, and who was called by Professor Raymond Dart, discoverer of his skull and jaw in Africa, "The Missing Link." The skeleton of Edward Drinker Cope, a well-known early paleontologist, resided in a large cardboard carton. *Gigantopithecus bilospanensis* was represented by a mandible. *Sinanthropus pekinensis* and *Pithecanthropus*, both also known as *Homo erectus*, and many more were represented by plaster casts.

When Eiseley had lunch in his office with the Museum Director, Froelich Rainey, or with another guest, they would sometimes laughingly but affectionately toast these many ghosts. Eiseley, Curator of the Early Man department at the Museum, was the first president of the American Institute of Human Paleontology which, aided by the Wenner-Gren Foundation for Anthropological Research in New York, purchased for relocation at the University Museum of Pennsylvania the Barlow collection, an important English casting business. Since the casting business included correspondence between Barlow and many well-known paleoanthropologists concerning important fossils of early man being discovered in Asia, South Africa and Europe, new ghosts added to the crowd in Eiseley's office: Franz Weidenreich, Davidson Black, Louis B. Leakey, Eugene Dubois, G. H. R. von Koenigswald, and Robert Broom.

Eiseley was well aware that man would have to propel himself down "many roads among the stars . . . in search of the final secret" of life, and although "the journey is difficult, immense, at times impossible, yet that will not deter some of us from attempting it." As all of Eiseley's readers are well aware, Eiseley himself was not deterred from this marvelous search: "Forward and backward I have gone, and for me it has been an immense journey."

It was a journey, as his readers are happily aware, into worlds of wonder, worlds from which all of us came and to which we might never have been able to return without the backward-in-time reminiscences of the great literary magician, Loren Eiseley.

It is easy to remember that the seven letter word "Eiseley" means the seven letter word "Enchant." It should be thus listed in all dictionaries, along with "Bewitch," "Charm," "Delight."

Naomi Brill

Loren Eiseley and the Human Condition

In the introduction to *The Star Thrower*, W. H. Auden wrote, "I suspect Dr. Eiseley of being a melancholic. He recognized that man is the only creature who speaks personally, works and prays, but nowhere does he say that man is the only creature who laughs."

In 1969 Loren Eiseley delivered a lecture entitled "How Life Became Natural," which was later published in *The Firmament of Time*. In it he tells of being locked, "in the evening twilight," in a museum hall filled with Crustacea. Gradually, as he waited to be released, he became overawed by the displays, particularly a huge Japanese crab, "one of the stilt-walkers of the nightmare deeps, with a body the size of a human head carried tiptoe on three foot legs like firetongs."

He concluded,

> "It is not the individual that matters, it is the Plan and the incredible potentialities within it. The Forms within the Form are endless and their emergence into time is endless. I leaned there, gazing at that monster from whom the other forms seemed flowing, like the last vertebrate on a world whose sun was dying. It was plain they wanted the planet and meant to have it. One could feel that massed threat of them in this hall.
>
> "It looks alive, Doctor," said the guard at my elbow.
>
> "Davis," I said with relief, "you're a vertebrate. I never appreciated it before but I do now. You're a vertebrate, and whatever else you are, or will be, you'll never be like that thing in there. Never in ten million years. Just . . . remember that we're both vertebrates and we've got to stick together. Keep an eye on them now, all of them. I'll spell you in the morning."
>
> Davis did something then that restored my confidence in

man. He laughed and touched my shoulder lightly. I have never heard a crab laugh and I never expect to hear one. It is not in the pattern of the arthropods."

Being aware of the uniqueness and healing nature of human laughter, even the recipient of its benison, does not mean that one has the ability to experience honest mirth as a saving reaction to the tragedies of life. A popular writer of the early twenties, in one of the immortal opening sentences of literature, wrote, "He was born with a gift of laughter and a sense that the world was mad. And that was all his patrimony." In many ways that is all the patrimony anyone needs in order to come to terms with life as it is and remain relatively sane.

But Eiseley was either not so endowed or else lost that ability somewhere in his journey through the years that were so painful to him. His writing is marked by his profound distrust of man as a species. It is the rare reference containing a positive comment about *homo sapiens* that is not at once balanced by an ominous negative one. Eiseley was a scientist and a keen observer so that he was full aware of human potential for good and evil and he conscientiously attempted to balance them in recounting human experiences but his terminology, the personal experiences he chose to share with his readers, is filled with darkness, terror, foreboding and a feeling of being constantly threatened. He, in a sense, described himself in each such episode and ascribed his feeling to all. In *The Star Thrower*, he wrote, "Our minds are haunted by night terrors that arise from the subterranean domain of racial and private memories . . . we inhabit a spiritual twilight on this planet."

Loren Eiseley was one of those seeming changelings that appear in the most unlikely places. Possessed of the inherent ability to see things differently and to acquire the necessary intellectual discipline, he was able to express his visions and ideas in ways that have significant meaning for a broad spectrum of readers. From the little we know about the development of creative intelligence it is not necessarily fostered by the smothering atmosphere of a conforming life, by protection from experiences that raise the challenging questions. The very stresses of his life contributed to his ability to realize his great gifts.

He recognized this when in *All the Strange Hours*, musing on his creative life, he wrote of "my unhappy parents' part in this dubious creation of a writer," "my father who heard and spoke beautifully," "my not so sainted mother who had left me the eye enjoying view." In *The Night Country*, speaking of "the beauty and terror of our mortal existence" he wrote, "I paid a certain price for my survival and indeed have been paying for it ever since. Yet the curious thing is that I survived, and, looking back, I have a growing feeling that the experience was good for me. I learned something from it . . ."

As an adult Eiseley could not but know that he was a truly remarkable man, but also a selected accidental combination of all that had gone before, his parents, his grandparents, even the "mad Shepherds," his forebears, as well as his life experiences. He knew that he was not totally self made, that from that household and the little world around it came something that freed him while at the same time left him so "scarred" that it colored the entire tenor of his creation.

Like all unhappy children, Eiseley felt that he was different from others, set apart by that "silent, unhappy household," by poverty, by illness, by isolation. The greatest responsibility for the "scarring" of his brain, which he barely survived, "which left me walking under the street lamps of unnumbered nights," he blamed on his deaf mother. In the final pages of *All the Strange Hours,* he wrote, "Her whole paranoid existence from the time of my childhood had been spent in the deliberate distortion and exploitation of the world around her."

In one of his epigrams, Oscar Wilde wrote, "Children begin by loving their parents. After a time they judge them. Rarely, if ever, do they forgive them," but forgiveness is essential or we can never forgive ourselves or others. Even with his participation in the field of medicine at a time when psychiatry and understanding of the human psyche were burgeoning; even with his knowledge of the terrible frustration of a person shut off from the world in childhood (his own temporary deafness as an adult surely must have sensitized him); even with his understanding of the fulfillment that came from the realization of his own creative ability and his knowledge of his mother's frustration; even though he knew that he was capable of those essential, sustained human relationships without which life can be meaningless – with his colleagues, his friends, his teachers, his wife, while "she quarreled and fought with everyone"; even though he knew worldwide recognition and financial security, while she knew only poverty and dependence on outside help, Eiseley was never able to forgive his mother. In the end he felt that her behavior, so destructive to him and to others, was deliberate.

Occasionally in his writings he voiced compassion, but some hurts, particularly those inflicted in childhood, are so deep and traumatic they cannot be healed. According to Eiseley, "great love bound those three (father, mother and child) together," but it is very probable that the "petulant violence" that caused his mother to beat the handle off her silver mirror, was expressed in violence against the terrified child himself. Certainly this reaction was what he expected and found in the world. In two instances – with the childhood gang leader who tormented him and the trainman who tried to kill him – he turned against them in a fury. In the latter instance, an old hobo comforting the battered young wanderer said, "Men beat men, that's all. That's all there is," reinforcing his previously held feeling that human violence was universal and to be expected.

But perhaps the greatest hurt, and one of which Eiseley apparently was not consciously aware, was inflicted by his father, who laid upon the terrified child the impossible task of being responsible for his terrifying mother. Such demand, common in such households, places a child in an irresoluble situation, for when he fails, as Eiseley did in that painful episode where he joins with the gang in mocking his mother, the failure leaves a residue of shame, guilt and resentment toward the loved parent. "And in the end," he wrote, "I broke my father's injunction." Is this perhaps why he could not sleep after his father's death? Why he needed human companionship and a light in the dark? Why he was haunted by recurring nightmares all of his life in which he strove to reach his father who "had an air of trouble about him," in order to learn something unspoken?

Carrying his scars with him, Eiseley went forth into the outside world in spite of his fearful mother's attempts to hold him close, to protect him from the life that had so hurt her. There he found rejection as all people do, but also an amazing amount of warmth and support that persisted throughout his lifetime.

There was his grandmother who helped him bake his clay heads in the oven in spite of her expressed reservations. ("them's no ordinary heads . . . they got that Darwin look."); there was Rat, a skinny, eager, tough boy who invited him into a gang, introduced him to the joys of the sewers of Lincoln, "created the world I live in, but died and left me in it"; there was his half brother Leo, to whom, Eiseley felt, their father turned in preference on his death-bed, but who brought him a copy of Robinson Crusoe and opened the world of literature to him; his Uncle Buck who introduced him to museums and supplied essential financial help in the years before he found his work; there was the English professor whose wealthy friend provided him a home in the desert when he had tuberculosis; there were friends, colleagues, teachers, people who promoted his career, the casual strangers he met in his wanderings.

An emotionally needy, suspicious person, always anticipating the slight, characterized by personal reserve and conscious distancing, Eiseley was still capable of reaching out to people and they responded to him and his writing. Some, like his father, recognized him as "a genius but moody"; some related to the suffering common to themselves as well as to him. In spite of his reservations and mistrust, people found warmth and concern as well as insight and beauty in what he said and wrote.

Eiseley described himself, as did *The Star Thrower* with whom he identified, as one who "loved not man but life." But man is part of life and you cannot love one without loving the other. For all his violence and cruelty, for all the unique brain that makes him different, makes him capable of attempting to control his world, often in a destructive way, man is only one of the Forms evolving according to what Eiseley called the Plan. Eiseley transcended his personal pain, grief, anger, bitterness, loneliness, terror and mistrust to reach a higher morality when he wrote in *The Unexpected Universe*, humankind "walks in his mind from birth to death the long, resounding shores of endless disillusionment . . . but out of such desolation emerges the awesome freedom to choose," and from that freedom comes "the expression of love beyond the species boundary."

In *The Star Thrower,* questioning his very ability to love, he said, "I love the small ones, the things beaten in the strangling surf, the bird singing which flies and falls and is not seen again. I love the lost ones, the failures of the world," and to these he turned for subjects. In some of the most beautiful and significant of his writings, in the only way he was able, Eiseley gave eloquent testimony to this love using the small, hurt creatures as protagonists, but what he wrote is equally true of other Forms, of people.

In *The Innocent Fox*, he wrote of the small fox pup that invited him to play, even as a human child might. In it he saw a happy world, "the universe . . . swinging in some fantastic fashion around to present its face, and the face was so small that the universe itself was laughing."

In *The Bird and the Machine*, he wrote about relationships, about the living versus the mechanical, through a pair of kestrels; he had captured the male for a zoo. As he removed the bird from the cage "I saw him look that last look beyond me into a sky so full of light that I could not follow his gaze . . . I just reached over and laid the hawk on the grass. In the next second . . . he was gone . . . straight into that towering emptiness of light and crystal . . . then from far up somewhere a cry came ringing down [from its mate waiting above]. The machine does not bleed, ache, hang for hours in the empty sky in a torment of hope to learn the fate of another machine, nor does it cry out with joy or dance in the air with the fierce passion of a bird" – and a human being! In his autobiography, Eiseley was able to write of his wife's closeness to him saying that between them "there was something too deep to explain." As they left his mother's funeral together he told her "you are the only one who knew everything, accepted everything."

Along with a feeling of exultation at the wonder of all life to be derived from reading Eiseley's work there comes a pervasive sadness and pity for the gifted, haunted man who carried such a heavy burden of terror, sadness, anger and guilt throughout his life. Could he have been so sensitive, written so well without them?

One of Eiseley's colleagues in the early days at the University of Nebraska told of having shared long walks with him during which they recited from and laughed over the writing of Don Marquis, the creator of Archy the cockroach, who wrote by banging his head on keys of a typewriter. His subject was Mehitabel, a gaunt, scarred cat from the alleys of New York City and her survivor's philosophy. She was the epitome of sardonic mirth and valiant understanding and acceptance of life as it is. Hers was the saving laughter of the suffering, the disposessed of the world.

If Eiseley could once honestly laugh with Mehitabel, he somewhere lost what might have been his salvation. In a sense he was one of the walking wounded of the world whose sensitivity and profound hurt are so great they cannot discover the peace they so endlessly seek. Although from the depths of his pain came writing so beautiful it spoke directly to the human condition and has a lasting place in the world's literature, would that he could have echoed Mehitabel not only when she said resignedly, "life is just one damn kitten after another," but also *"Toujours gai!"*

University of Nebraska Press

Also of Interest:

ALL THE STRANGE HOURS
The Excavation of a Life
By Loren Eiseley
Introduction by Kathleen A. Boardman

In *All the Strange Hours*, Loren Eiseley turns his considerable powers of reflection and discovery on his own life to weave a compelling story, related with the modesty, grace, and keen eye for a telling anecdote that distinguish his work. His story begins with his childhood experiences as a sickly afterthought, weighed down by the loveless union of his parents. From there he traces the odyssey that led to his search for early postglacial man—and into inspiriting philosophical territory—culminating in his uneasy achievement of world renown. Eiseley crafts an absorbing self-portrait of a man who has thought deeply about his place in society as well as humanity's place in the natural world.

ISBN: 978-0-8032-6741-1 (paper)

THE FIRMAMENT OF TIME
By Loren Eiseley
Introduction by Gary Holthaus

Loren Eiseley examines what we as a species have become in the late twentieth century. His illuminating and accessible discussion is a characteristically skillful and compelling synthesis of hard scientific theory, factual evidence, personal anecdotes, haunting reflection, and poetic prose.

ISBN: 978-0-8032-6739-8 (paper)

THE LOST NOTEBOOKS OF LOREN EISELEY
By Loren Eiseley
Edited and with a reminiscence by Kenneth Heuer
Sketches by Leslie Morrill

This indispensable collection is filled with marvelous autobiographical glimpses of Loren Eiseley at different points in his life: as a young, inquisitive man during the Depression, as an astute archaeologist, as a blossoming writer, and, lastly, as a world-renowned observer and essayist. Also included are poems, short stories, an array of Eiseley's absorbing observations on the natural world, and his always startling reflections on the nature and future of humankind and the universe.

ISBN: 978-0-8032-6747-3 (paper)

Order online at www.nebraskapress.unl.edu or call 1-800-755-1105.
Mention the code "BOFOX" to receive a 20% discount